THE Independent Inventor's Handbook

THE
Independent
Inventor's
Handbook

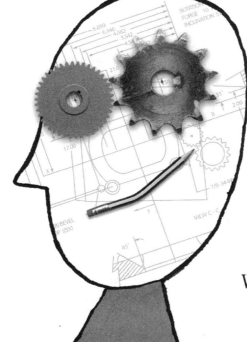

BY
Louis J.
Foreman
AND
Jill Gilbert
Welytok

Workman Publishing
New York

Library of Congress Cataloging-in-Publication Data is available.

ISBN 978-0-7611-4947-7

Workman books are available at special discounts when purchased in bulk for premiums and sales promotions as well as for fund-raising or educational use. Special editions or book excerpts can also be created to specification. For details, contact the Special Sales Director at the address below.

Workman Publishing Company, Inc.
225 Varick Street
New York, NY 10014
www.workman.com

Printed in the United States of America

First printing April 2009

10 9 8 7 6 5 4 3 2 1

Contents

The Entrepreneur & the Attorney

From the Entrepreneur

Everybody has a great idea, and that great idea has the potential to change the way we live, the way we work, or the way we play. The problem is that most people never follow through on making their idea a marketable reality. It could be that you don't have the financial resources—what, you're not willing to take out a second mortgage on the house or risk your kid's college tuition to pursue an idea? Maybe you don't want to leave your day job—you've got a great one and leaving it to pursue your invention is too big a risk. Or perhaps you don't completely understand the process of having an idea and making money from it.

So how, exactly, *does* the average person transform a mere sketch on a napkin into a useful product and make it available to the masses? Single mother and secretary Bette Nesmith Graham did it when she turned her 1956 idea of "mistake out" (now known as Liquid Paper or "white out") into a million-dollar business by 1967. And seventeen-year-old Chester Greenwood went from grammar-school dropout to millionaire godfather of earmuffs. Band-Aids, Frisbees, zippers, the Scrabble board game, steam irons, Lifesavers candies, online bookstores, and bagless vacuum cleaners—these are all examples of massively successful patented products that started as humble but brilliant ideas. And the list of United States patents is more than 7 million strong.

So what's the secret? The plain answer is: There isn't one, but these inventions are linked by the fact that they began as unique ideas, they served (or continue to serve) an identified consumer "want," and they target a specific and ready market. These ideas were also tirelessly championed by their "parents"—the inventors who created them. Jeff Bezos, founder of Amazon.com, told me the story of how he borrowed a pickup truck to drive across country and start his online bookstore because he didn't want to be sitting back when he was eighty years old wondering what his life would be like if he had actually pursued an entrepreneurial dream. "Most regrets are acts of omission rather than commission," he says; so he just did it. I recently met James Dyson, whose story began with a prototype for a bagless vacuum cleaner crafted from cereal boxes and duct tape and continues, 5,127 prototypes later, with the number one selling vacuum cleaner in the world and more than $1 billion in sales.

But enough about their stories, because, as inspiring as they are, this book is really about yours.

For whatever reason, the vast majority of great ideas never surface. And Jill, my coauthor, and I would like to change that. We've set out to provide you with all the tools you need to educate yourself about the invention development process. We've paired these lessons with examples of individual inventors who have done it before, so you know it's possible for you—yes, you—to turn your idea into a successful, marketable product.

I like to think of pursuing an invention or entrepreneurial endeavor as a journey. Like a traveler planning a trip, you need to know how much it's going to cost to get to your destination before you start driving. Many times I see inventors and entrepreneurs who are so excited about the prospect of making a fortune from their idea that they forget about what it's going to take to get there. They forget to gas up. Or they plan for only one fill-up along the way. Going forward with an idea without a plan for paying the costs of bringing your invention to the market is the biggest mistake you can make; and it's often what stalls people. Making money off a great idea requires incredible patience and an investment of time and financial resources—all topped off with a bit of luck.

My own entrepreneurial journey started as a child: I was always coming up with creative ways to make money, whether it was running a backyard carnival or selling greeting cards door to door. By the time I reached college, my mother's dream of my becoming a doctor was well off course (the "C" in organic chemistry should have been a clue). During my sophomore year, I started running a small T-shirt company out of my fraternity room. A few short years later, it had grown into one of the largest screen-printing and embroidery companies in the United States, with more than 150 employees and nearly 80,000 square feet of manufacturing space.

I sold that company in 1995 and moved to Charlotte, North Carolina, to pursue another business opportunity: starting a NASCAR apparel company called Track Gear. Fortuitously, Track Gear went from zero in revenue to over $15 million in sales in less than two years; I sold that business to a publicly traded company in 1997.

A little more than ten years out of college, having become a bit of a serial inventor, I reached a goal that many entrepreneurs hope to achieve: With five successful startups under my belt, I had set myself up for early retirement.

But that entrepreneurial spirit just wouldn't go away, and I realized very quickly that I wanted to keep working in business, only now with the intent of helping others do what I did.

I've spent the last decade mentoring hopeful inventors and entrepreneurs and holding small-business seminars for groups at community colleges. It's exciting to sit with a group of aspiring entrepreneurs and offer advice—sometimes providing people the inspiration to move forward and sometimes offering a healthy dose of reality to help them understand that the timing isn't always right and reassessment may be in order. And I founded Enventys, a unique industrial-design and marketing firm that helps develop and launch products for top companies in the United States.

The dot-com decade spawned a stellar lot of billionaires, but my journey wasn't built on computer coding or visions of technology. My own products and inventions have always been about people and how we live our everyday lives. I work with what I know—many of my inventions and companies have been related to serving just about every sport for which I'm a spectator or participant, including a line of NASCAR clothing and a high-performing soccer shin guard. My vision is not about computer screens; rather, it's about the evolving consumer market—that beating pulse of our culture and economy.

Over the years, I came to realize that as exciting to me as these one-on-one mentoring conversations and talks to an audience of thirty to fifty people were, there is a vast number of people out there who would never hear the words I was speaking. So when the opportunity came for me and my company, Enventys, to be involved in the development of the PBS television program *Everyday Edisons* (a reality show on which individual inventors receive help in developing their ideas into actual marketable products), I jumped at the chance. I wanted to spread the message that I have come to know so well: If you are committed to getting your product out there, and if there is an actual demand for that product, you can be successful. And that is really the inspiration behind *Everyday Edisons*—and this book in your hands.

Every inventor's path is unique (after all, the invention he or she makes must by its very nature be new and unique), but it's easy to get caught up in thinking that the same rules apply to everyone. Between kitchen appliances, car parts, board games, sports equipment, medical tools, odd engineering mechanisms, and just about anything else you can imagine, the possibilities for new products for invention are endless. The paths to producing them are just as varied, and for this, inventors need advice. At our first *Everyday Edisons* casting call, hundreds of inventors showed up even before the studio opened, lining up for hours for a chance to show their inventions and receive some direction.

This handbook is here to help: To empower you, as an individual inventor, to understand the value of innovation and embrace it wholly, to prepare for the journey ahead, and to make reaching the destination of success that much more enjoyable.

You'll realize you can do many things that you'll learn from this book yourself, but I hope there are also things you'll see you can delegate or ask for help with. Resources are all around you. There are inventor groups online that can help with support; there are online communities such as Edison Nation (www.edison nation.com) that provide a community of inventors to turn to and resources that can help you avoid making the classic mistakes. There are professionals—patent attorneys, design firms, engineering firms, branding and marketing companies—who do this every single day and can help you with the process. And then, of course, you can, and should, surround yourself with people who can provide you with support—and not just the financial kind. The key to success as an individual inventor is to surround yourself with people who complement your skills and who believe in your vision: people who have your best interests in mind and can answer the questions to which you don't know the answers. In short, good road buddies, because it is a long journey and you've got to make sure you're not making the trip alone.

Over the years I've met with CEOs, R&D leaders, the head of Hewlett-Packard's HP labs, and others to learn about tomorrow's technology. What that technology will bring to us is absolutely amazing, but I'm most excited about the innovation that will surface as a result of *Everyday Edisons* and of this book. As an inventor, I've been there, and I've helped other people get there, and I wish I'd had this book along the way to help them get their ideas out in the world and do it in a sensible and productive way. The idea that you can is what gives me hope. Hope that future generations of inventors will continue to make the United States a country that is built on a foundation of innovation.

From the Attorney

My dad was an independent inventor. He invented toys and novelty items, and made a living selling them through ads in the back of magazines (which was where people sold such things before the Internet). My dad's most notable invention was a realistic looking fish, mounted on a plaque, that could suddenly start wiggling. Since no one had seen anything like it at the time, there was that element of surprise that made it an instant hit. He had other products, too, like plants that started dancing and a scale that talked to you. They were all part of his "It's Alive!" product line.

Unfortunately, my dad never applied for a patent on his fish invention. A skeptical patent attorney advised him against it, not thinking the fish was as humorous as we all did. But my dad had clearly discovered a niche: Over the next thirty years, our family watched as a curious trend swept the planet. An international market developed for the "It's Alive!" fish. Generations of gyrating, singing variations hit store shelves. But because my dad had never filed for a patent, the idea was in the public domain, free for anyone who wanted to copy and develop it.

In fact, one factory that my dad had hired to manufacture his model had sold his original mold to competitors. The fish was a financial bust for our family and a market coup for countless other companies. And it certainly made an impact on me. I had witnessed the "It's Alive!" concept move from idea to product to market phenomenon, and its success had a profound effect.

Today, as a licensed patent attorney, my corporate clients pay me well, but my true love in this profession is helping individual inventors bring their ideas to fruition in the marketplace.

Inventors from all walks of life come to me in hopes of hitting it big with their inventions. With raw ideas for everything from kitchen gadgets to thermonuclear devices, these inventors are pumped up to file patents on their inventions almost as soon as an idea pops into their heads. That enthusiasm is an absolute necessity in the business of innovation, but I try to keep people from filing patents on inventions unless they have the ability to bring them to market. After all, taking an invention from idea to product is a time- and money-intensive process—one that an inventor has to be prepared for. Usually people are grateful for my honesty, but one particular day when my phone was ringing off the hook, I was getting nothing but arguments from would-be inventors who claimed that my legal advice contradicted what they had seen on a new-at-the-time, sensational network show called *American Inventor*.

I welcome any call from a hopeful inventor, but I didn't like seeing individual inventors being misled on national television, especially by a show carrying the label "reality series." While *American Inventor* was arguably entertaining, the commercials for "invention submission" companies that flood the late-night airwaves are a huge problem in the inventing community—and far from amusing. Every year, these fly-by-night companies with seemingly endless advertising budgets bilk young inventors out of millions of dollars by telling every inventor that they have a fabulous idea and will make a fortune. These companies charge inflated rates for worthless patents and never actually bring anything to market. Sadly, I have seen more than one aspiring inventor burn through a retirement fund and end up with nothing to show but a slick-looking binder of worthless documents from one of these spurious companies.

Needless to say, when I first saw the e-mail from a local chamber of commerce announcing a "casting call" for inventors for a new television show, I was pretty cynical, and my scam alert was at code orange. But I was pleasantly surprised when I clicked through the link for the *Everyday Edisons* show: The first thing that struck me was that it was produced by a successful and innovative product development firm, Enventys, which has plenty of Fortune 100 clients and a huge list of successful products and companies that are household names. The creator, Louis Foreman, was a frequent speaker at the U.S. Patent and Trademark Office Individual Inventor conferences. And most amazingly, the objective of the show seemed to be to educate inventors, and even pay them a 10 percent royalty for the rights to their invention. It looked as if the show's producers were actually going to make the investment to turn a raw idea into the product. This was not an *American Inventor* copycat. This was inspiring.

I have wanted to write a book for inventors for a long time. Shortly after another *American Inventor* episode aired (followed by the predictable and exhausting deluge of cold calls and e-mails), I contacted Louis with this idea: that successful inventorship could be taught in a book. He called me back within the hour, enthusiastic and ready to collaborate.

This is that book. We've combined my expertise as a patent attorney and Louis's expertise as an entrepreneur and inventor to create what we believe is the complete guide for the individual inventor—a collection of all the advice, based on our experience working with more than one thousand inventors, that we use to help our clients take their ideas from sketches on napkins to products on shelves.

In the pages that follow, we've shared what you need to know in order to sift through your dreams to decide which are worth pursuing in the real world and how to make your ideas succeed. This book is a testimonial to what individual inventors can accomplish, and *how* they (you!) can attain those goals. These are lessons that any inventor, no matter the medium, can adapt to suit the needs of any idea—even if your development studio and center of operations happens to double as your kitchen, your college dorm room, your garage, your nursery, or your backyard.

You, as an inventor, have the gift of imagination. From your head to paper to prototype to market, here's the resource that will help you take your invention to the next level.

CHAPTER 1

Eureka!

Turn an Invention into a Venture

The fact that some inventions thrive while others languish isn't about lack of capital or lack of connections. The success or failure of a good idea depends on the vision and approach of the inventor.

Know that you, a lone inventor in the vast galaxy of the marketplace, can gain control of the information you need for determining the fate of your invention. By approaching the market systematically (and without too much financial investment), you can take your invention to the next level.

The Sport of Inventing

Bringing an invention to market is one of the toughest challenges you can take on, and to succeed, you must enjoy it as a type of intellectual sport. Along the way, you'll have to alter the habits and mindsets of your customers and weigh the risks in your own life. The process is rigorous and demanding and requires significant patience, strategy, discipline, and endurance. If you love the process of inventing, and you approach it with objectivity and creativity, you're bound to succeed sooner rather than later, and without draining your savings or risking your future. It's a matter of sifting through those imaginings and deciding which are worth pursuing in the real world.

The Independent Inventor's Handbook is your personal coach as you fight to bring your idea safely and successfully into the national or global arena. It will teach you how to evaluate ideas for their profit potential, how to work toward developing them in such a way that others can share your vision, and how to steer clear of scams and financial pitfalls. The rest is up to you.

Although you must have a genuine affinity for the inventive process, it's alright to view it as a path to profit as well. A properly developed and executed idea *should* turn a profit for you. It should also give you critical experience that can help you bring a future inven-

tion or product to market. As you'll learn, inventing and product development go hand in hand and are virtually inseparable processes.

The Difference Between an Idea and an Invention

This book is about how to turn an *idea* into an *invention*. There is a vast difference between the two. An idea is an unproven concept that is a product of your imagination. It pops into your head but, in that raw state, is far from something anyone can implement or profit from. Ideas are rarely, if ever, marketable commodities until they become inventions. If you want to have something to sell, you *must* maneuver through market research, product development, compiling cost information, and in all likelihood, securing a patent. As this book will explain, it is only after you have done these things that you have a licensable or sellable product. Raw, untested ideas are rarely a commodity for which anyone will pay.

Lots of people can and do make their living (and even their fortune) by inventing. But it is a long road from your brain to the bottom line, and you can't expect to take a mere concept and license it for "a million dollars." Inventing is work, and it's coupled with an element of risk.

Obstacles All Inventors Face

Whether you are a housewife, a scientist, an engineer, or a student, the obstacles you'll face are relatively universal. Here are a few of the realities you'll have to be prepared to take on as you join the ranks of inventors who have helped humanity evolve from apes to astronauts.

■ **Skeptics:** You, like all inventors, will encounter skeptics. Be they cynical coworkers, corporate higher-ups who don't want to hear about competing products, or just the folks you share beer with every other week, they will find their way into your life. Don't be discouraged by the naysayers; they can, in fact, be very valuable assets: Their uninformed perceptions are sometimes the source of insight.

Don't ignore their barbs and jealousies. Analyze the underlying objections instead, so you are prepared to respond well when you put your invention in front of the people who have the power to advance it.

■ **Teaching against the times:** Most great inventions teach something that contradicts that which is already "known" or accepted. Clerics disputed Galileo's concept that the Earth was round (see profile, next page). People will dispute whether you have defined a problem that is worth solving, or whether your solution will win acceptance. In most cases, you can simply

Tuning In: The *Everyday Edisons* Show

Ordinary People, Extraordinary Ideas

Fortunately, you can now turn on your television or computer and get accurate information about the inventing process. Each week, the *Everyday Edisons* show appears on public television, chronicling the progress of a group of ten to twelve real inventors with real ideas as they protect their ideas, turn them into inventions, and actually get them into the consumer marketplace. Coauthor Louis Foreman's company, Enventys, is a product design and engineering firm that attracts many inventors. In an effort to better educate and inspire individuals with great ideas, he created the concept of *Everyday Edisons* and turned the idea into an Emmy award–winning show. The show works with the cooperation of the U.S. Patent and Trademark Office, which has taken a special interest in the program and has had its top officials appear on it.

The *Everyday Edisons* panel of judges includes product-development and patent-law experts who look for ordinary people with big ideas. Inventions can be in any stage of development, ranging from sketched concepts to patented designs and prototypes. To watch episodes of the show online or to find out when a casting call is coming to a city near you, check out the show's website at www.everydayedisons.com.

take your own more current market, manufacturing, and technical research and step around those who are stuck in the past.

■ **Finding the funds:** Yes, it costs money to build prototypes, protect concepts with patents, and create awareness of your ultimate product. You may well need to invest your own funds at the start. Once you

can prove market acceptance, others will step forward to help you underwrite production, and friends, family, and strangers may be persuaded to invest, as well. But no one will (or should) invest in a concept for which you have not done preliminary groundwork. Once you can prove that a market exists, and that you have found a cost-effective way to proceed, purchase orders or licensing agreements will come your way. And investors will want to hear what you have to say.

■ **Abandoning ideas that are not worth developing:** People who become overly invested in ideas, despite all the evidence in the world that there is no market for them, cannot succeed. If your idea falls into this category, you must be willing to abandon it and pursue the right idea at the right time. No amount of hard work or investment capital can turn a trivial problem into a consumer need, or make licensees and retailers enamored with a product that appears to be unique to your tastes and personal experience. If you are an inventor with an active, creative mind, then other, workable concepts will come to you. You must leave yourself free to pursue the ones that are perceived as having merit in the marketplace.

When they are bringing a new product to market, all inventors ultimately follow certain universal principles of inventive success. These basic marketing principles have not only withstood the test of time, they have made the

INVENTOR PROFILE
Galileo Galilei:
Teaching Against the Tide

Mathematician, physicist, astronomer, philosopher, inventor . . . Galileo Galilei (1564–1642) was one of the most brilliant, visionary, innovative men in Western history. What was his secret? For one thing, he regularly challenged popular beliefs; in fact, he often ridiculed others' incorrect ideas (which didn't help him win any popularity contests during his lifetime). Galileo challenged the widespread notion that the Sun revolved around the Earth, advocating the Copernican heliocentric (sun-centered) view instead. This didn't sit well with the Roman Catholic Church, since Galileo's views flatly contradicted its interpretations of Scripture passages; he was ordered not to hold or defend this heretical idea, and later was placed under house arrest and forced to recant publicly.

Galileo also happened to be a genius at combining pure theory with applied science. While he didn't actually invent the first telescope, his improvements transformed the device into one of the most important instruments of the Scientific Revolution, and the telescope led to his own major astronomical discoveries, including sunspots, the rings of Saturn, and the four large moons of Jupiter.

United States the most successful innovation nation in the world.

Inventors share one important trait: vision. But being visionary doesn't guarantee that you will attain financial success. Take Galileo, for example, one of the most famous individual inventors of all time. Many of the tools he invented are still used today—the thermometer, for instance, the compass, and the most famous: based on his vision that the Earth was round and not flat, the telescope. Sadly, Galileo couldn't gain market acceptance for his "round Earth" theory; he later recanted his hypothesis, and there is no evidence that anyone ever bought a single telescope from him.

The world has come a long way since the sixteenth century. Today there is a global marketplace in which large companies compete for new products and ideas from individual inventors, with billions of potential customers and niche markets linked by the Internet. Today individual inventors such as Galileo are revered. And for the past two centuries, the United States (which didn't even exist in Galileo's day) has led the world in innovation, in large part because of its patent system. As President Abraham Lincoln said, the system rewards inventors by adding "the fuel of interest to the fire of genius."

The United States patent system gives inventors "exclusive rights . . . to their discoveries." It's a system designed to empower individual inventors like you. Your ideas are your

FROM THE ENTREPRENEUR
BIG BUSINESSES START SMALL

Small businesses with innovative products and services create 80 percent of new jobs in the U.S. A young business with a successful invention doesn't require a factory and hundreds of employees at first—sometimes ambition alone is enough to get you started. Google started out as a dorm-room experiment; Ben and Jerry learned the ice-cream trade from a correspondence class; my own first business began in a college fraternity room.

From a $5 course to a multi-million dollar franchise.

"intellectual property," and they are constitutionally protected so that you may license this property to others for a profit or use it to build your own business. Once your patent is issued, it precludes others from making, selling, or using your invention for twenty years. Initial patent protection (called a provisional patent application) generally costs only a couple thousand dollars, which keeps the system both affordable and accessible to just about everyone.

The Business of Creativity

Every great idea needs to be commercialized to see the light of day, and this requires that you focus on the problems of the actual user in the marketplace. To be honest, most new products or inventions that take the market by storm are not things that most consumers actually *need*. For the most part, they have everything that's necessary for their lives. But consumers *want* to purchase new things that are perceived as better or may make their lives easier or more enjoyable. A successful invention is something users need or want that is offered at a price they are able and willing to pay—and not every idea *can* be brought to market with a realistic price. This is particularly true in recessionary times. Consumers need to feel that their expenditures are logical and that their wants can be justified as *need*. When economic times are tough, certain businesses, such as grocery stores or pharmacies, do better (people forgo big expenses and rationalize treating themselves to necessities or smaller luxuries).

One of the toughest parts of the process is recognizing when to abandon ideas, and when to pursue them doggedly, directing your energy toward working through each problem as it arises in the manufacturing and marketing process. The ability to solve problems, absorb information, and let go of ideas and assumptions in the face of information, and the desire to seek that information, are the basis of a successful inventive mindset.

Though there are many one-hit wonders in the inventing world, most inventors are serial inventors: They work on many ideas before they find success. Failure and abandonment are inevitable parts of the inventing process. Thomas Alva Edison held patents for 1,093 inventions. Not all of them were successful. For example, one vision of his that did not endure was that of using

Unable to find buyers for his concrete furniture, Thomas Edison found success building bigger structures.

concrete for building furniture. Edison was so dogmatic in his belief that cement was the way of the future that he formed the Edison Portland Cement Co. and invested in concrete prototypes for cabinets, furniture, pianos, and even phonographs. Unfortunately, his concrete products were financial disasters. Users didn't want to sit on concrete chairs, nor did they want to pay for them. The important part is that Edison learned from his failure: He revised his idea to make it successful. If concrete could not be used for smaller structures, he'd go for larger ones instead. Concrete produced by his company was used to build Yankee Stadium in the Bronx—something consumers did want and for which they were willing to pay the necessary price (at the ticket counter, the concession stand, et cetera). The stadium remained more than eighty-five years until the new Yankee Stadium was built in the Bronx in 2009.

Too often inventors make the fatal mistake of clinging to a single idea in the face of all market evidence to the contrary. For example, one inventor, whom we'll call Gus, was convinced that consumers would pay for a specially designed umbrella to position above the barbecue grill when it rained. Several people pointed out that the device added significant cost to the grill, and that even though it prevented the food from getting wet, the person who was grilling would still get soaked! Despite this feedback from these potential customers, Gus proceeded with a costly grill umbrella prototype and began soliciting "investors" to help him bring his invention to market. For all we know, he is still out there, bitterly complaining that if he could only raise the capital, he could get his product to store shelves.

In contrast, consider the case of George Crum, the commonly credited inventor of the potato chip. Back in 1853 Crum was a chef at the Moon Lake Lodge resort in Saratoga Springs, New York, where French fries were a restaurant menu staple. One day a diner complained that his signature fried potatoes were too thick. Crum began experimenting with making thinner and thinner fries, eliciting feedback from his customers with every batch he served and revising his process accordingly. Crum had the inventive mindset we hope you'll have. A combination of trial and error, continuous consumer feedback, and revisions to the recipe led to the invention of the potato chip. Although Crum never got a patent on potato chips, his invention was a market success. And although he could not foreclose his competitors with a patent, he did eventually open his own restaurant, which boasted clientele that included William and Cornelius Vanderbilt, Jay Gould, and Henry Hilton. Today the potato chip business is a $6-billion-a-year industry and employs more than 65,000 people.

Other inventive "potato heads" admirably followed in Mr. Crum's footsteps, finding further success by keeping their eye on the potential consumer.

A customer complaint turned into a $6-billion-a-year industry.

(continued on page 10)

INVENTOR PROFILE
George Beauchamp:
Launching an Invention in Tough Times

George Beauchamp (1899–1941), the inventor of the electric guitar, was a typical serial inventor. He had varied interests, his dining room table was his laboratory, and his process was not without challenges. His most famous idea, the electric guitar (eventually distributed under the name Rickenbacker), for which he applied but never gained patent rights, was launched during the Great Depression and initially met with skepticism from musicians.

Over the years, Beauchamp obtained a number of patents on musical instruments. His patents included the steel guitar pick (U.S. Patent No. 1,787,136); a guitar with a steel body (the now-famous "Dobro") for playing "slide" guitar (U.S. Patent No. 1,808,756); and the lap steel guitar, nicknamed the "frying pan guitar" (U.S. Patent No. 2,089,171, shown here).

With a passion for music and a knack for science, Beauchamp used the then ordinary knowledge that metal moving through current could be translated into an electric current by a nearby coil of wire. He was determined to make a prototype (a model on which the final form of the product would be based) that applied this knowledge to the guitar. After many months of trial and error, and using his washing machine's motor to wind the electric coil, he enlisted a friend to help him use a sewing machine motor.

In the early 1930s Beauchamp launched the Electro String Instrument Corporation (which is still in existence) with a friend, guitar craftsman Adolph Rickenbacker. Beauchamp contributed electrical and amplification concepts, and Rickenbacker created the body design. The company began production in a small rented facility next to Rickenbacker's tool and die plant. The two received modest initial investments from family and musician friends (much the way we advise doing in Chapter 7).

But the timing could not have been worse. The Great Depression was at its lowest depths, and to make matters worse, the United States Patent Office delayed issuance of a patent for more than three years as it attempted to sort out whether the "frying pan" guitar described in the patent was an electrical device or a musical instrument, since no patent category covered both. In the meantime, competing companies rushed to get an electric guitar onto the market. By 1937, when the patent was finally issued, it seemed futile for the small company to wage legal battle against all the potential patent infringers.

Nevertheless, the Rickenbacker guitar gained market popularity and a loyal following among musicians. In particular, the artist Les Paul contributed to the recognition of the guitar by using it when he performed. Eventually, Electro String introduced fully electric bass violins, cellos, and violas, inspiring and liberating musical instrument designers for generations to come. When Beauchamp died in 1941, his funeral procession was over two miles long and included many great guitarists of the day.

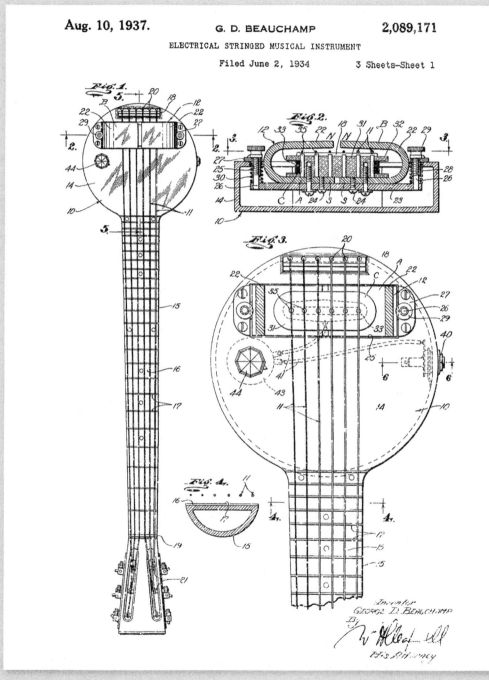

Aug. 10, 1937.   G. D. BEAUCHAMP   2,089,171

ELECTRICAL STRINGED MUSICAL INSTRUMENT

Filed June 2, 1934   3 Sheets—Sheet 1

Electrical device or musical instrument? The "frying pan" guitar initially defied patent category.

(continued from page 7)

Until 1926 chips were dispensed from barrels. A former nurse named Laura Scudder developed the sealed paper bag as an alternative, thus making wider distribution possible. She continued to modify her design, based on customer feedback, as to the most effective types of paper for keeping chips fresh and preventing oil from seeping through. Ultimately she moved from plain paper to wax-lined bags. In 1955, the business having survived the Great Depression, Laura Scudder, Inc., was sold to Borden, a large food company, to the tune of $55 million.

The Inventor's Curse: Too Much Cash Too Soon

Who said "It takes money to make money"? There are many inventors out there who know this isn't always true. Money is certainly necessary to bring an invention to market, but what it really takes to be a successful inventor is an objective and a solid understanding of your target market.

Money can help you at the right stages of the inventing process. However, too much spending too soon, before you have a handle on the market, can be a disaster. One inventor (we'll call him Kyle) was an engineer by trade and by no means wealthy. He took out a $50,000 second mortgage on his house while he worked on his blinking bicycle wheel prototype. He fastened magnets to the spokes of the bike wheels to cause an electrical current that could power a set of flashing lights on the wheel. Kyle

spent months poring over drawings and selecting materials to make the perfect prototype and apply for a patent. Through various contacts, he was able to present his patent pending product to several large bicycle manufacturers. The results were uniformly disappointing.

Kyle soon learned, through his contacts, that the companies he'd approached had previously considered and rejected similar concepts. (Since they'd decided not to go forward with the idea in the early stages, they had never sought patents, so there was no public information to discourage Kyle from pursuing the idea.) The companies' marketing departments thought that Kyle's price point was too high for a novelty product for the casual cyclist, and that serious cyclists (racing or touring) wouldn't want colorful flashing lights.

The companies had no interest in carrying something that looked cool but would most likely be rejected by their consumers. The unfortunate thing is that Kyle knew many reasons why cyclists would use his wheels: Parents would view his product as a safety enhancement for their kids' bikes, and adult cyclists would use it to make themselves visible at night. But rather than convey any of these key pitch points about his target market to the bike companies, he spent his valuable presentation time showing them how perfectly his prototype worked. He not only lost the $50,000 he invested, but he also lost confidence in his idea. And as far as we know, Kyle never pursued another invention.

Market research should always come before the patent and the prototype. There is no economic justification for a patent that doesn't protect a market niche or for a working prototype of a product no one wants to buy. Kyle's scenario might have been different. If the bank had turned him down for a mortgage, for instance, he would have had no choice but to begin with market research. He could have presented the companies with a working but less polished prototype to demonstrate how his wheel functioned, along with sketches to show how he imagined the finished product would look. Kyle might have come prepared to demonstrate that there was indeed a "blinking wheels" market and he could have financed his invention himself rather than borrowing against his house prematurely. There is a time and place to take this risk, but not before validating the existence of a market.

Bootstrapping Your Invention

In the mid-1800s there was a popular American author named Horatio Alger who wrote more than a hundred dime novels that were characterized as "rags to riches" stories. Alger's novels were bestsellers in their time. The public loved reading the formulaic and inspiring plots about down-and-out characters who attained wealth and success through their own hard work.

We like to think of every individual, or independent, inventor as a sort of Horatio Alger protagonist, a person who is capable of starting with next to nothing in the bank at the beginning of the story and financing their invention on their own, i.e., "bootstrapping" the invention into store-shelf successes.

There are a few ways bootstrapping

Retailers and distributors are always looking for new products to stock their shelves.

your invention can work for you that we'll discuss throughout this book. The first is by going to a customer and getting a significant advance commitment (or purchase order) for your product. You can do this once you know when and how you'll manufacture your product and how much you'll need to charge. Retailers and distributors are always looking for new products to stock their shelves. A purchase order is a contractual commitment from the company to purchase a certain quantity of your product, at a determined price, to be delivered within a specified time. You can turn this purchase order into power.

The important point here is that a purchase order, especially one from a major retailer, is a bankable asset. You can borrow against it from a bank or from a *factor,* a company that finances projects like this on a short-term basis (for a higher interest rate than banks charge). You can also take your purchase order to a manufacturer and convince the company to do a *joint venture* with you.

For example, if the manufacturer normally requires payment in net 30 days (net amount due 30 days from receipt of goods or service), you can negotiate for terms of net 90 days, which will allow you to pay off the manufacturer with the money you get from your retail customer. If all goes well, you collect your payment at the end of the process, and bank your profits after everyone else is paid off.

We talk about how to get a handle on manufacturing costs in Chapter 3, and how to go for the purchase order in Chapter 7. For now, you just need to know that this strategy exists and both individual inventors and big companies use it to work together all the time.

LEVERAGING BY LICENSING Another great way to bootstrap your invention is by licensing it to someone else (and letting them worry about procuring the purchase orders). When you license your invention, you're not responsible for manufacturing or selling it. This means there's a lower risk for you, but usually there's a lower return, too, than if you sell it yourself. The advantage of licensing is that it requires very little cash outlay and minimizes many of the risks. John Suckow from Milwaukee learned to leverage his inventions in exactly this way.

In 1994 Suckow was running his floor-installation business when he had the inspiration for his first invention: a new tool that could make laying (and stretching) carpet a lot easier. Laying carpet was a tedious process which until then required the installer to return to the site and restretch the carpet after a few months. Suckow knew that his solution (leveraging the applied force more efficiently onto the carpet rather than back into the power stretcher) had merit, so he made a prototype of the tool on his own and tested it on the job. When it worked as planned, he decided it was time to take his invention to market.

Fortunately, Suckow did not have a lot of spare income at that point, or he might have blown it paying one of the invention submission companies that advertise on late-night TV to market his invention for him. Instead, he went to his local library in search of information. The reference librarian on duty directed him to the *Thomas Register* directories, which contain several volumes about products and the companies that make them. Suckow located the top four tooling manufacturers in the United States and wrote each of them about his carpet stabilizing product. All four turned him down.

Suckow felt discouraged, but he swallowed the rejection and decided, after applying for a provisional patent application to protect his invention, to send one of the companies a prototype. The prototype was rejected, too, and was returned to him; it didn't even look like the package had been opened.

After thinking about the situation for a few more days, Suckow called the president of the company, using the contact

US007114704B1

(12) **United States Patent**   (10) **Patent No.:**   **US 7,114,704 B1**
Suckow   (45) **Date of Patent:**   Oct. 3, 2006

(54) **STABILIZING DEVICE FOR A CARPET STRETCHER**

(76) Inventor: **John G Suckow**, 1513 Lake Dr., Hubertus, WI (US) 53033

(*) Notice: Subject to any disclaimer, the term of this patent is extended or adjusted under 35 U.S.C. 154(b) by 0 days.

(21) Appl. No.: **10/908,171**

(22) Filed: **Apr. 29, 2005**

(51) **Int. Cl.**
B25B 25/00 (2006.01)
(52) **U.S. Cl.** **254/209**; 254/200; 294/8.6
(58) **Field of Classification Search** 254/200–212; 294/8.6
See application file for complete search history.

(56) **References Cited**

U.S. PATENT DOCUMENTS

444,415 A *	1/1891	Anderson	....	254/211
3,963,216 A	6/1976	Victor	....	254/62
4,008,879 A	2/1977	Youngman	....	254/57
4,230,302 A	10/1980	Crain, Jr.	....	254/212
4,750,226 A *	6/1988	Costill	....	7/103
4,815,708 A	3/1989	Samson	....	254/212
5,129,696 A	7/1992	Underwood	....	294/8
5,145,225 A	9/1992	Muller	....	294/8
5,150,884 A	9/1992	Hyer	....	254/209
5,183,238 A	2/1993	Sorensen	....	254/209
5,364,143 A	11/1994	Grady	....	294/8
5,607,141 A	3/1997	Clark	....	254/200
5,681,031 A	10/1997	Foley	....	254/209
5,765,808 A *	6/1998	Butschbacher et al.	.....	254/200
5,931,447 A *	8/1999	Butschbacher et al.	.....	254/200
5,984,274 A	11/1999	Medwin	....	254/200
6,161,818 A	12/2000	Medwin	....	254/200
6,170,612 B1 *	1/2001	Krumbeck	....	187/200
6,371,446 B1	4/2002	Gauthier et al.	....	254/201
6,669,174 B1	12/2003	Vito	....	254/212

* cited by examiner

Primary Examiner—Emmanuel M Marcelo
(74) *Attorney, Agent, or Firm*—John K. McCormick

(57)   **ABSTRACT**

A carpet stretcher is provided with the improvement of a stabilizing apparatus and method that attaches to a carpet stretching tool to prevent the carpet stretching tool and any of its additional stretcher tubes from bending or bowing when utilized to stretch carpeting during carpet installation. The stabilizing apparatus consists of a bar with appropriate means for attachment to a carpet stretcher.

7 Claims, 6 Drawing Sheets

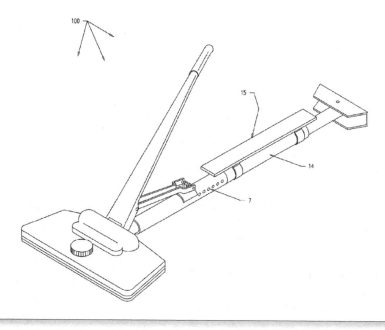

John Suckow's carpet stabilizing device, patent no. 7,114,704.

information from the *Thomas Register* directories. The company president was brief, but he listened. At the end of the conversation, he told Suckow to send the prototype once again, this time directly to him. Within the week, Suckow got an excited call—the company had been trying for more than twenty years to figure out a way to solve the problem.

Today Suckow's carpet installation tool, the Stabilizer, is sold worldwide, and a royalty check arrives quarterly. He now holds patents or patent applications on more than thirty other products, including a Laser Compass, a high-powered repelling magnet, a windshield cleaning apparatus, and other outdoor items, many of which are sold in major retail outlets such as Wal-Mart, Gander Mountain, REI, and Blain's Farm & Fleet.

If John Suckow had paid an invention submission company, or delegated the marketing to someone who didn't know the flooring market as well as he did, he'd still be laying floors for a living. Instead, he did his own market research and leveraged his knowledge of that market into a license with a simple working prototype that he made himself. (We'll cover licensing in a lot more detail in Chapter 6.)

Checklist for Success: Taking on the Marketplace

It is vision, not venture capital, that makes or breaks an invention. It is the ability to see a market for a product that does not yet exist and to confront and overcomes obstacles as they arise. There are certain common denominators in getting inventions to market. Regardless of whether yours is a thermonuclear device, a software product, or a kitchen gadget, it must have the following qualities in order to succeed:

- ☑ Creativity
- ☑ Marketability
- ☑ Credibility
- ☑ Patentability

You'll be surprised at how naturally each of these elements comes to you if you truly have the heart and the mind of an inventor. We are not asking you to reinvent yourself; we are asking you to use the capabilities that you already have in order to give your good ideas the chance they deserve.

FROM THE LAWYER
THE VALUE OF A PATENT

 A patent legally prevents competitors from marketing your novel product— which is why a patent is only valuable for products for which a market exists. It's impossible for inventors to bring all of their inventions to market. Remember that you'll always have more ideas, so wait for a strong one with solid market potential before you devote time and money to turning the idea into a product.

Your Innate Creativity

An inventor who is just starting out needs to understand that creativity has nothing to do with education or IQ. Creativity is inherent and will not leave you, so don't worry that your first idea will be your last. Inventive success is much more of a continuous process than a single "Eureka!" moment. Nearly every unworkable idea can be recycled or reused as valuable information to feed the next great idea, in order to bring just the right products to market at just the right place, time, and price.

Profitable and, innovative companies know that they must have an ongoing supply of creativity from independent inventors like you to continue to come up with the new products they need to serve the marketplace. Procter & Gamble, the world's largest consumer products company, has set a goal of acquiring 50 percent of its new products from ideas outside the company walls. A. G. Lafley, chairman of Procter & Gamble, told *Fortune* magazine in 2004 that "we're just as likely to find an invention in a garage as in our labs."

Procter & Gamble is by no means alone in its outlook. Other large companies, from all sectors of retail and industry, have the same desire to find innovative ideas outside of their company walls. Many companies will not hesitate to open their doors to credible inventors who approach them with a basic prototype, solid market data, and preliminary patent protection in hand.

Identifying Your Customer

It's not enough to have a vision of something that works great. You can sit around all day thinking about how to perfect your prototype, but to take your invention out of your basement or garage to market, you need to refine the design and function to appeal to an *actual* customer. This means identifying who your customers will be and talking with your potential customers about your product.

Before you can do this, you need to paint a picture of those customers. How old are they? What's their education level? What is their marital status? How much do they earn? Where do they live? Do they have kids? There are two reasons for doing this. The first is to target specifically the buyers who constitute your market, and the second is to be able to determine the number of customers in that market. Once you identify who the customers are, you can ask them about your product.

What kind of dialogue should you have with your prospective user?

The questions on the following two pages are a place to start. Let them be your guide as you work to identify and clarify the needs and wants of the potential audience for your product. Feel free to photocopy the pages and use them like a workbook to take notes each time you interview a member of your target market.

"Where will you go to buy my invention?"

_____

_____

_____

_____

"Will my product appeal to you on sight, or will you need to know more about it to understand its value? At what point will you make your decision?"

_____

_____

_____

_____

"Why will you see my product as a solution to your problem, when there may be many other ways to do it, not to mention you've lived your entire life without my product?"

_____

_____

_____

_____

"If you don't actually need my product, what will motivate you to buy it?"

_____

_____

_____

_____

_____

"How would you describe yourself? Are you young? Middle-aged? Athletic? A single parent? Do you have a special hobby or interest that will draw you to my product? Do you have children, pets . . . ? What kinds of items do you buy regularly and where do you buy them?"

_____

_____

_____

_____

_____

"If you like my product, how will you pay for it?"

_____

_____

_____

_____

CASE STUDY

Inventive Inspiration from Friends and Hobbies

Best pals and fellow scrapbookers Mary LaValley, Pam Hester, and Deborah Mance originally met through their children, but it was their love of scrapbooking that drew them together and kept them in close contact over the years. By the time they came up with the idea for Arccivo, the women knew the characteristics of their potential users as well as they knew one another. They were, after all, members of their own target demographic: middle-aged moms whose hobby time was limited but important—important enough that they would pay for a product that solved the common scrapbookers' challenge of transporting the precious projects from one place to another and keeping them intact.

The trio's remedy was to create an elegant, portfolio-shaped case that opens like a book, provides two work surfaces for craft layouts, and can easily be moved from one place to another without causing items to shift. On the *Everyday Edisons* show, viewers watched the transformation of the rough prototype into a must-have tool for scrapbooking wizards and novices alike. Removable magnetic sheets covered and protected unfinished pages, locking loose paper and photos in place. Case zippers made it easy to transport in-progress pages safely and securely.

Sure, at $59.95, Arccivo's price can be considered a splurge. But these inventors knew their niche market (within a $3 billion industry) would bear it. Not only did they understand the mindset of a hobbyist, they realized that their imaginary user was likely to come from a dual-income family, and that working moms especially would welcome a device that helped them make the most of their hobby time.

After you've had a conversation with your prospective user, it's time to have a talk with yourself. Make what you've learned visible: write lists and notes, run a spreadsheet. You need to know how much it will cost to make your product, who will make it, and where the money will come from to tide you over until you get that first purchase order. The sooner you figure out these essential, basic questions, the better.

You also need to be able to advocate your user's needs, so that if you end up pitching your invention to a company that shows interest in licensing your idea, it feels as though you're speaking on behalf of a dear friend. This is how it was for inventors Mary LaValley, Pam

Hestor, and Deborah Mance, who developed the Arccivo, a new patent-pending scrapbooking portfolio featured as one of the star inventions on *Everyday Edisons* (see Case Study, previous page).

Give Credibility to Your Claims that the Product Will Sell

Many products on store shelves sell well based on their package and design, but a product's merits need to be more than skin deep. Companies won't invest in something that simply looks cool; it has to have legitimate sales potential. Whether you're trying to negotiate a license or secure a purchase order, you need to be credible enough to convince the company that your product will sell. You can gain this credibility in two ways: proof of actual sales or believable market research.

GOOD THINGS COME IN SMALL PURCHASE ORDERS Past product sales are generally a pretty good indicator of future product sales, at least in the short term. (This is why Fortune 500 companies like to purchase existing companies with a track record of profitable sales.)

As an inventor, getting national distribution with a mass merchant should not be your biggest worry. A series of small-scale product sales can lead, like a trail of bread crumbs, to lucrative deals. Small orders at trade shows, from local businesses, and over the Internet translate into dollars as well as data (which is more important when you're starting

FROM THE ENTREPRENEUR
LEAVING "EVERYONE" BEHIND

By limiting your target market, you have a better chance of 1) reaching that market and 2) reaching beyond it. It's impossible for your audience to be "everyone"—the sooner you identify the ways in which your potential consumer is distinct, the more successful your product will be.

out). These venues are all experiments to credibly demonstrate how your product will sell in a much larger market, given the right exposure and a shot at mass distribution.

SAYING IT WITH STATISTICS Don't underestimate your computer as a research tool. Describing your intended users is one thing; pointing to statistics that prove they actually exist is another. Chapter 2 tells you how to do market research in more detail. For now, what you really need to know is that all kinds of direct and indirect data you uncover will work, so long as it truly indicates reasons your product will sell. For example, one pair of inventors showed up at the *Everyday Edisons* casting call with a bizarre device for disposing of chicken bones without having to leave the table (the tray holds chicken wings and lets you slip the mess of the leftover bones out of sight through a hole in the center). Though their prototype looked a bit like a science-fair volcano, the *Everyday Edisons* judges were easily convinced of the quirky invention's value when they learned that 10.5 billion wings were sold in 2004 alone (see Case Study, next page).

Losing Credibility by Claiming Too Much

There is one answer companies don't want to hear when they ask you the question "Who are your customers? Who will buy your product?" That answer is "Everyone." If you claim that your market is everyone, it only shows that you haven't done your homework. Your market, logically, cannot be everyone. There are some products, from toothbrushes to televisions, that nearly everyone does buy. But even within these broad categories, there are hundreds and sometimes thousands of distinct brands, styles, and designs, each directed at a particular subset of customers with the unique characteristics that drive their daily choices and make them choose one toothbrush or TV set over another. Why would consumers suddenly purchase your product when they've been using something else or living without it?

Take the garage door opener, for example: It's reasonable to assume initially that every car-owning household would want one, or that every household would at least benefit from having one, but it's unlikely that drivers in an urban center even have the luxury of using individual garages. So your audience must own not only at least one car but a garage as well (therefore, nearly all of them must live in a suburban or rural location). Who

---

CASE STUDY
Getting Numbers that Prove Your Niche

Brent Anderson and Russ Stanziale arrived at their idea for a new and improved way of serving chicken wings while they were eating at their favorite wing joint one evening. As they observed the "disgusting pile of bones" their feast produced, the wheels began to turn. They followed through on their creative concept for a new serving platter that would conceal the debris, and presented their papier-mâché model at the Atlanta, Georgia, casting call for *Everyday Edisons* in July 2005.

Despite the obviously handmade prototype, it was the pair's clear knowledge of the size and personality of their potential market that convinced the judges. The chicken wing industry was booming, there was a need to be met, and the price point (at $14.99) was right.

The SnacDaddy platter targets an impressive market.

is your target driver? Male or female? A business person? A carpooling parent? What kind of car do they drive? Narrow down your universe of target consumers by asking additional questions.

Or take something as simple as sliced bread, everyone's favorite invention. It can be argued that everyone would want to have sliced bread, but in reality, everyone is not the target purchaser of the product. If you are a manufacturer of sliced bread, you need to determine who is consuming your product. Is it kids? Adults? Both? More important, who purchases the product: moms? dads? other adults?

Brand, price point, and characteristics (ingredients or components, packaging, and so on) all influence the consumer's decision to purchase bread, and further segregate the consumers who will opt to buy your brand of sliced bread instead of some other brand.

As much as inventors would like to believe that everyone will buy their product, the truth is that the market is much smaller than they think, and they need to convince the largest segment of the market that *this* product is a good fit for them.

Among users, your target market is the subset who will become choosers of your product. It is your job to define the subset convincingly and accurately.

Patentability

Companies don't license raw ideas; they license patented inventions. Regardless of what you may have heard, patents are critical to getting an invention to market and are probably a lot less expensive to obtain and maintain than you may have been led to believe. (See Chapter 8 to learn how to get a patent.) There are three very logical reasons companies want you to have a patent or a patent pending before you talk to them. First, a company doesn't want to risk being sued by a third party waving a patent and claiming that the company is stocking its store shelves with an infringing product that you sold to it. Larger companies don't have time to do a patent search on every product that's brought to them, so a patent operates as a sort of certification that this research has already been done by the United States Patent and Trademark Office (USPTO). A pending application does not carry such a certification, but may include an opinion letter from an attorney on the likelihood of patentabililty. A "patent pending" designation is also a signal that you understand how the patent system works, and are smart enough to avoid intentionally infringing upon someone else's patent.

Not only are companies worried about lawsuits by third parties, but they're also worried about a lawsuit by you. They want to know exactly what the invention is that you're bringing them, and they want to know that the scope of that invention is defined by the claims contained in your patent application. Many companies are worried about "contamination," an industry term for a

The inventor's dream: to invent the best thing since sliced bread.

Inventors Who Are Celebrities and Celebrities Who Are Inventors

When we talk about inventors, Thomas Edison is usually the first person who comes to mind. During his life, Edison achieved the status of celebrity inventor. After all, he earned 1,093 U.S. patents throughout his career and was responsible for some of the most important inventions in history. Imagine life today without the lightbulb, recorded music, or movies.

Today, the concept of the celebrity inventor is quite different. People still become famous through inventing, but more often today we see celebrities who have achieved their fame in other areas who have also come up with some pretty ingenious ideas. Here are a few of these "celebrity inventors" and the ideas they came up with.

Marlon Brando: An invention to tune a drum (Patent No. 6,812,392).

Harry Connick Jr.: A system for displaying music for an orchestra that replaces sheet music (6,348,648).

Harry Houdini: An improved diver's suit that is easier to put on and take off (1,370,316).

Michael Jackson: Method and means for creating antigravity illusion (5,255,452).

Hedy Lamarr: Secret communications system (2,292,387).

Abraham Lincoln: Method for buoying vessels over shoals (6,469).

George Lucas: Multiple design patents on *Star Wars* characters and vehicles.

Prince: The ornamental design for a portable electronic keyboard musical instrument (D349,127).

Mark Twain: Patents on garments, games, and improved scrapbooks (140,245; 121,992; 324,535).

Eddie Van Halen: Musical instrument support (4,656,917).

situation in which they've been working on an idea in-house prior to meeting with an inventor, and the inventor later sues them for stealing his or her idea. These lawsuits are costly, and they're terrible for PR. If your idea is patented, the issue of what you actually brought to the company, and at what point in their own process, is conveniently resolved by looking at the claims in your patent application. Companies worry about this contamination issue so much that many will return your proposal unopened or even refuse delivery.

The last and most important reason for getting a patent is that it gives you a huge competitive advantage in the marketplace. Though patents are not easy to obtain, they are what the framers of our Constitution intended you to have. It is possible to get certain short-lived products to market without a patent, but doing so means you must be prepared to fend off the competition by constantly cutting your price or increasing your advertising. You have to compete because you don't have a patent to act as a barrier to their market entry. A patent acts as a twenty-year mini monopoly on what you've created. A pending patent is like posting a Keep Out sign on your intellectual property. Competitors looking to enter your market space will see your patent application as a legal obstacle, which they must overcome by designing around your product or by waging legal warfare. If you have broadly drafted patent claims, your competitors' clumsy attempts to circumvent them may actually undermine the marketable features of their product. Since all patent applications and issued patents are ultimately published on the Internet at www.uspto.gov, your would-be competitors will see yours, and you can see theirs.

CHAPTER 2

First Define the Market,

Then Design the Product

In 1921 Earle Dickson, a cotton buyer for Johnson & Johnson, noticed that his wife, Josephine, kept cutting her fingers when she was preparing food. To wrap her wounds, she often made herself bandages from cotton gauze and adhesive tape. The tape wasn't very effective, though (it wasn't designed to stick to skin), and the bandages themselves were bulky and troublesome to make. Armed with the image of a real consumer and an everyday problem, Dickson began a series of experiments that led to the invention of the Band-Aid brand bandage.

When Dickson's boss at Johnson & Johnson saw the early prototype of the Band-Aid, he was thrilled. He decided to bring it to market, and even made Dickson vice president of Johnson & Johnson for bringing him the idea. After all, the potential audience was huge—who hasn't cut a finger? But here comes a lesson in the realities of inventing. During the initial years, Band-Aid sales were surprisingly *slow*. Dickson's vision of the user (housewives) was sound, but he met with an obstacle most inventors face: Consumers are often reluctant to change their way of doing things, even when a better way is offered. Housewife consumers were not seeking a solution to a problem they felt they already had solved: They had tape and gauze in their homes. Why should they adopt a more expensive solution? And what was the incentive for a retailer to carry a new product when consumer demand hadn't been created? Though the solution to the problem wasn't as simple or as elegant as a self-adhesive bandage, it was an existing solution nonetheless. It wasn't until Johnson & Johnson made the marketing decision to give free Band-Aids to Boy Scout troops as a publicity stunt that sales soared. These impressionable miniconsumers took the product home and asked their parents to buy them. Excited children who had no reservations about trying something new were just the push their parents (and

Before the Band-Aid, gauze and tape were the accepted solution for bandaging scrapes and cuts.

Johnson & Johnson) needed. The rest, as they say, is inventor history.

The "Why Bother?" Question

There is a big difference between needs and wants. Arguably, as consumers, we have most of the things we need to survive. We justify buying new products because, as human beings, we want to improve our existence and bring beauty (as we each perceive it) into our lives. Still, just like the Band-Aid, every product must overcome the "Why bother?" question.

As an inventor offering a new product to the market, your greatest challenge will be to convince people to try your product or method, and to offer it at a price they will think is worth paying. At the start, ask yourself: "Who will buy this product?" and "Why will they bother to try it?" Even if your product is better and cheaper, even if it solves a problem with which consumers have struggled for years, you must still convince them to add something new to their lives.

Every consumer of your product must become properly acquainted with it first, so that they feel it's a logical choice and can make an affirmative decision to purchase it. Sometimes people are convinced of their needs and wants through advertising campaigns. Take, for instance, the one for George Foreman's famous "lean mean grilling machine."

The ads for this electric skillet made people believe grilled beef prepared in this device is a more healthful food, making it a logical choice.

At other times consumers are actively seeking solutions to problems, as was the case with Dr. William Scholl and his insoles. In 1904, at the age of twenty-two, William Scholl patented an arch support insert that he called a Foot Eazer. He developed the product to address the foot pain that affected many people at the time. As an apprentice shoe repairer and then a shoe sales clerk, Scholl was inspired by his customers and pursued a medical degree to study the structure of the foot, which led to the invention of the Foot Eazer. After his first invention, he developed additional new products such as callus pads, cushioned insoles, and orthopedic shoes. He is credited with creating the over-the-counter foot-care category by addressing a consumer problem and inventing an effective solution to eliminate the pain.

Once consumers are convinced that they have an unmet need, they must be further persuaded that their problem is solved by your invention, and it must be offered to them at a price they're willing to pay. Moreover, consumers are extraordinarily fickle: They change their purchasing habits as soon as they are confronted by new advertising messages or budgetary realities.

For instance, instead of relying on faxes or the postal delivery system to send messages, consumers can now choose to access the Internet via Treo, BlackBerry, iPhone, or at a Wi-Fi Internet café. They cast their consumer votes with dollars. Not only do they constantly make choices, they may change their minds at any time. The capriciousness of consumers, and their ability to be swayed, creates enormous opportunities for inventors because companies must always strive to anticipate what today's consumers will find attractive tomorrow.

For individual inventors, as well, the dynamics of the consumer market are continually evolving. People's basic needs are very similar, but their wants change. We did not need TV before it was invented, but once it was, most of us wanted it. The VCR gave us more uses for our TVs, and once it was introduced, it quickly became the most wanted home entertainment product. Regardless of changes in technology, the consumer will always determine the success or failure of products because, ultimately, a product must address a need as it is perceived by the consumer and be designed with the consumer in mind. What you're looking for is a product that consumers believe they "must" have: a product they want to own because it makes their life easier or more enjoyable or one they just aspire to own. Even industry titans such as Google and Microsoft are constantly competing to educate their consumers enough to convince them to try something new. Individual inventors have the same challenge: to convince consumers

When the BlackBerry offered instant e-mailing on the go, consumers needed little persuasion to purchase the product.

to take the risk of changing their behavior in some way.

For years, Microsoft has dominated the market in productivity software, selling hundreds of millions of copies of its office software. Its model of selling an application that is loaded on a com-

A niche market is based on two elements: an actual user with an actual need not being adequately met by the mainstream market.

puter's hard drive has served Microsoft well, allowing it to become one of the world's largest and most profitable companies.

Could there be a better way? Google, which revolutionized online advertising, is attempting to prove that there is. Its approach is the use of Internet-based software that need not be installed at all. Software that is purchased on a CD is expensive. Businesses must hire a certain percentage of their workforce just to deal with security issues and viruses, software conflicts, bugs, patches, updates, and myriad other issues that are frustrating to every business. Currently these problems are solved independently by the IT people in every workplace. Google figures there is a better way.

Google wants to convince workers to connect to software using the Internet so that it can address their software and maintenance issues all at the same time. Using a Google Apps application

instead of a Microsoft Word program, users can create documents by simply logging on to the Internet instead of using programs that are loaded in their computers. Innovations such as faster computers and high-speed Internet make this possible. Google argues that if businesses move to accessing software from the Internet, "there's no hardware or software to install or download" and "there's minimal setup and minimal maintenance." But Microsoft has a 90 percent market share, which means that 90 percent of users are trained to use Microsoft programs (Word, Excel, and so on). Thus, even if Google Apps were free (which is not the case for large companies), there will be a cost to the employer for training workers to switch to it. The cost is in educating its employees (users) to try something new, and in using a new solution to meet their needs. The question is, is Internet-based software *really* a better way?

Developing products that consumers want and trust can be accomplished by design and experimentation. Battles of innovation, such as those between Microsoft and Google, are what have made the U.S. economy so prosperous. But it is important to remember that consumer needs and wants are not always productive, functional, or even logical. Sometimes we just "want" a product that looks better or is more in tune with our thoughts, tastes, and surroundings or that we find entertaining (such as a new game or music CD). The desire to express and differentiate ourselves through our purchasing choices (for instance, buying an iPhone versus a BlackBerry, or wearing Nike versus Adidas) is an equally valid consumer need.

Tapping Into Niche Markets

Most of us are familiar with the term "niche" or "target" market. Most companies whose products are household names know that a key to success is to find this sort of specialized market in which to launch a product. Many start by appealing to the needs or wants of niche customers and then increase their distribution and sales as the market becomes mainstream.

As discussed in Chapter 1, if you are asked to define your niche market when you make your pitch, the worst answer you can give is "everybody." No matter how promising the statistics are, not everyone will buy a product, no matter what that product is. The "everybody" answer only shows potential partners that you haven't done your homework.

The constant need to find new market niches is why companies have huge research and development departments and why they are willing to open their doors to individual inventors. (See Chapter 5 to learn more about how to be one of those inventors who makes it through the door.) New ideas can create new market niches.

As Earle Dickson and his Band-Aid so aptly illustrate, a niche market is based on two elements: 1) an actual user with 2) an actual "need" that is not being adequately met by the mainstream market. Companies regularly hire legions of market researchers, industrial designers, sociologists, engineers, analysts, and other professionals to identify consumer needs and then design products to fill them more effectively.

Even if these teams of researchers worked 24/7, however, their reach would still be limited, because they can never fully know what is going on in the millions of homes and workplaces across the country. While some research allows us to peer into some people's bedrooms and bathrooms and offices, there is no method that can totally reveal how human subgroups act and think. And data about daily life spurs innovation. Professionals with substantial training may be immersed in product development and called upon to make

FROM THE LAWYER
KNOW YOUR POWER AS AN INDIVIDUAL

The power of the individual inventor is built into United States law, and it's been that way since the founding fathers drafted the Constitution. Article I, section 8 protects the intellectual property rights of all inventors, stating the role of government "To promote the Progress of Science and useful Arts, by securing for limited Times to Authors and Inventors the exclusive Right to their respective Writings and Discoveries."

critical decisions about which products are brought to market, but you, as an individual inventor, have the unique perspective and insight from your own job, home, and relationships that large companies value. They have a finan-

cial incentive to license ideas from you because you enable them to reach segments of the population that would otherwise be untapped. This is why individual inventors will always be important to our economy and to com-

CASE STUDY
Cultivate a Blue Sky Mindset

With more than twenty-one products under his belt, Joe Casale is what we consider a professional inventor. Before hanging out his shingle as a successful independent industrial designer, Joe was something of a corporate insider who spent a number of years working with (and learning from) major corporations in the toy industry.

In 2006 Casale, armed with just a sketch (albeit a highly detailed one), had one of his ideas selected for development on *Everyday Edisons*. It was an all-purpose, time-saving cutting board with measuring drawers and a built-in grater that allows users to measure as they chop, separate out the scraps, and cut down on clutter on the kitchen counter. While acknowledging that some of his more far-fetched ideas warrant working prototypes, Casale maintains that you don't necessarily need a prototype to get noticed—rather, he suggests, your presentation technique should be uniquely based on the best way you can convey your idea. Meaning that a sketch can be just as effective as a model in successfully licensing an invention if it fits the situation and the product.

Much of Casale's success as an inventor is attributable to his keen understanding of what customers want. He cites what he calls the

"blue sky" mindset: He stretches his creativity and imagination to the maximum in order to solve a market problem, but keeps his feet firmly on the ground because an idea needs more than creativity to be viable in the marketplace. "Everyone has an idea," he offers, "but the bottom line is, that idea will have to be bottled and packaged."

And as an independent inventor, you don't have to do it alone. Identify and take advantage of the resources, connections, and expert knowledge around you. Though you can initiate much of the process yourself, be ready to acknowledge when you've exhausted your area of expertise, so that you can seek professional help.

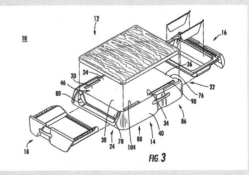

A patent drawing for the Koku cutting board from inventor Joe Casale's pending patent application.

panies that want to stay competitive in consumer markets.

Many inventors make a mistake in thinking that the appeal of a new product to a company is the device, software, or game itself, and that the goal is to sell the company on it. In reality, your goal as an inventor is to convince the company that you have identified a new niche market and to follow with an explanation of how your product fills it.

Prove Your Niche with Numbers

Proving the existence of a niche market is a lot like proving a scientific theory. In order to confirm theories of black holes, evolution, or even the existence of atoms and neutrons, scientists experiment and gather mathematical data that indirectly proves the existence of these things. Likewise, if you can gather enough indirect data to make it probable that your niche exists, you've proved your theory that there is a void in the universe of consumer products.

When you are doing research to demonstrate the existence of an untapped market for your invention, look for indirectly related facts that indicate the existence of actual people and communities with a need. For example, if you have invented a device to help tomato plants grow straight, you may want to find out how many tomato seed packets people buy every year. If you've invented an improved baby bottle, find out how many baby bottles are sold annually. If you are fine-tuning a device to help

teachers grade papers (see Case Study, page 35), look for statistics on the number of elementary, middle, and/or high schools in the United States, the number of licensed teachers in each state, or the number of students at various grade levels. All this data can help make the case that a consumer market for your invention actually exists.

The more reliable indicators of real potential consumers you can cite, and the more specific the data you have about them, the better prepared you are to prove your theory that there is a new niche market.

Needs and Niches Never Stand Still

Most marketing experts agree that you have identified a new market niche when you've identified an unmet "need," or a void, in a substantial enough segment of the population to justify creating a product for them. But this is an oversimplification.

Words like "need," "want," and "niche" are marketing shorthand for concepts that are more complex. You must recognize that, as a rule, consumers are already satisfying the need somehow. Maybe they are using reusable cloth diapers, when you want to sell them biodegradable synthetic ones; maybe they're addicted to using gauze and tape to patch their wounds when you are offering them a Band-Aid; or they're using a corded headset to talk

From a detailed sketch to product: The Koku cutting board.

hands-free when you want them to use a wireless Bluetooth device.

Regardless of the manner in which they are currently meeting their need, that's your competition. It doesn't matter if the present solution in no way resembles what you have to offer; it still constitutes competition. A train can get you to your destination; a plane can do it faster. A train doesn't look or operate much like a plane, but rail was the chief competition for the infant commercial aeronautics industry in the early 1900s. The competition for those little touch screens you electronically sign at the checkout counter of a store when you make a credit card purchase was once carbon paper. In its own way, it fulfilled the same "need," but who "wants" the carbon-coated receipt in their wallet?

Consumer "need" is a very subjective term, and the line between a need and a want is a blurry one. These terms are simply a way to describe a perceived willingness to change. We can probably all live our lives in relative comfort without using a newly invented product. But most people would not want to declare a moratorium on buying or trying new consumer products, even though it may be economically and environmentally sound to do so. In truth, they would see a dearth of new products as a dismal, depressing prospect in which we could not vary our surroundings and lives. We all have a willingness to change, but how much, when, and for what is the issue all inventors face in bringing their products to market. Consumer tastes will always be shifting as fast and as efficiently as communication can travel (which is faster than ever). Consumers will want new products that they see or hear about. Each new product they purchase solves a perceived need or want in a particular way at that point in time, but it creates new needs and wants and consumer expectations for additional products, features, conveniences, and aesthetic improvements. This is the dynamic, opportunity-filled universe in which inventors have always operated, and the universe that governs your fate as an inventor.

So, if consumers always want new products, why is it so hard to bring something new to the market? The answer is that consumers and companies have infinite choices for new products and they have a limited amount of money. (Those of you piping up about the Pet Rock, please hold that question—we'll get to it later in this chapter.) Your new cooking device not only competes with other cooking devices, but it also competes with the prospect of dining out. Inventors and entrepreneurs live in a constant state of "pick me." Retailers have only so much shelf space, and manufacturers can only fund the development of so many new products at a time. Consumers, as much as they want new products, are skeptical about parting with money when it involves a learning curve for every new product they see. You must demonstrate that a consumer will feel strongly enough to pick your product. You must also overcome consumers' resistance: They know there may be something else they will want to spend their money on tomorrow, and besides, today they already have an acceptable solution to what you are offering. There needs to be a "value proposition," evidence that the solution your product offers is worth the price. Your product has to be "cost justified" in meeting whatever need or want it was designed to respond to. The stronger the need and the greater the projected value proposed to the consumer, the better your odds of introducing it.

For example, let's take a look at how a creative little company named iRobot has anticipated a need for its product, the Roomba. The Roomba is a robotic vacuum cleaner that whisks around the user's house without their direct involvement. It senses walls and other obstacles and moves around them. Unlike traditional vacuum cleaners, the Roomba is a flat disk that ventures under beds and furniture to attack the dust bunnies that have been plaguing householders for centuries. Since

The Roomba, a robotic vacuum cleaner, convinced the market of its value in cleaning hard-to-reach dust bunnies.

2002, the company has released several versions of the product (including a variation that mops floors, as well), constantly revising it based on customer feedback.

But the Roomba comes at a cost. When it was first introduced, it was perhaps $100 to $200 more than a traditional vacuum cleaner with traditional attachments for getting under beds. And while the price continues to decrease as the technology becomes more affordable, the Roomba faces the same obstacle your product will face. Why should the consumer change? Why should a person use a Roomba when a traditional vacuum cleaner gets the job done? Why not spend the money on a very good digital camera instead? The inventors of Roomba bet that the smaller size (easier to stow), easy-to-use interface (aesthetically pleasing and functional display panel), and most important, labor-free aspect (any multitasker's dream) are enough to make consumers want to purchase their product. Because of these features (and a

---

You must think like a consumer. Be alert to potential problems and be able to address them.

---

dedicated marketing plan that includes in-store demo models that show just how "irresistible" the device is), iRobot is slowly but surely garnering a loyal client base. The company's no-questions-asked return policy shows confidence in the product and addresses consumer concerns about whether the Roomba may be less durable than a vacuum cleaner. Assuming the vacuuming capability is the same, the value proposition here is that of having the robotic device do the work for you.

Overcoming Resistance to Change

Thomas Jefferson allegedly once said, "If it is not going to pick my pocket or break my leg, why do I care?" Getting consumers to try any newly invented product usually involves helping them understand how to use the product, and requires an investment in the cost of the product. At a minimum, they have to figure out where to buy the product, and conclude that the cost (money and time spent) over the present solution is worth their

while. And consumers know that any technological device requires them to read directions. Sometimes the best inventions are the simplest to use because they eliminate the issue of the learning curve as an objection.

To address the potential obstacles to bringing your product to market, you must think like a consumer. Be alert to potential problems and be able to address them. If you don't communicate with consumers who draw your attention to your product's faults (all products have them), you'll never be able to see (and implement) the solutions.

Consider the example of Rebecca Byler, a career teacher who invented a pen called the Edugrader, featured on *Everyday Edisons*. The pen keeps track of the number of wrong answers on a test and then accurately displays the percentage score based on the total number of questions on the test. Because teachers have been correcting and grading papers in ink ever since paper became cheap to produce, Rebecca's job as an inventor was to convince teachers that using the Edugrader was worth both the out-of-pocket cost and the learning curve. The current standard is familiar but cumbersome: mark the wrong answers by hand, add them up, subtract that number from the total number of questions, then divide by the total number to get the correct percentage. Some tech-savvy teachers will adapt to and adopt the new product right out of the gate; others may be more resistant to the change. Ultimately, the "grades" these teachers give the Edugrader, and the subsequent incorporation of the teachers' feedback into the product design, will determine the product's initial success and its long-term acceptance.

Should Your Invention Be Developed?

Only the market can give you the answer to this question. Inventing can be a solitary act,

---

CASE STUDY

Inventing What You Know

A career teacher who's taught every age level over the course of twenty-seven years, Rebecca Byler admits that while her passion for teaching hasn't changed, the demands of teachers have. Piles of paperwork have kept this Nashville native—and countless other teachers—from focusing on what's really important in the classroom: teaching.

Frustrated by the amount of time she spent grading papers the old-fashioned way (using the old slide-rule grader that has been used for decades), Rebecca came up with the idea for an electronic pen that would keep track of right and wrong answers, as well as calculate the grade. So, with a rough prototype she made using a Sharpie marker, Byler was off to an *Everyday Edisons* casting call.

Design, branding, and engineering teams worked in unison on *Everyday Edisons* to transform Byler's rudimentary prototype into a sleek, high-tech new tool perfect for tired, time-crunching teachers looking for a way to lighten the grading load.

Teaching is a substantial part of Byler's life. As an inventor, she was able to make a good case that there was a niche market for the product she envisioned. Rebecca is just the type of inventor to whom companies must open their doors if they want to stay competitive. Her decades in the classroom gave her specialized knowledge in a field that no market research department could simulate. Although Byler's solution is a high-tech one, as opposed to Earle Dickson's Band-Aid decades ago, the principles of the marketplace aren't radically different across invention genres and generations. Like Dickson, Byler started by identifying a problem and then focused on inventing a solution. Getting the Boy Scouts on board with Band-Aids helped Dickson's invention break into the marketplace; similarly, Byler was able to garner national media attention through her appearance on television, which substantially helps the learning curve of the consuming public.

The Edugrader was invented by a teacher to help improve her craft.

FROM THE LAWYER
HIRING A MARKETING FIRM: YOUR RIGHTS UNDER THE INVENTOR'S PROTECTION ACT

Where do you turn if you're an inventor with a promising idea? Many companies may very enthusiastically offer to help you patent and market your invention. They all but promise to land you a lucrative licensing deal for your invention. It is understandably tempting to hire an "experienced" company to get a patent, a prototype, and "everything" you need. Unfortunately, many aspiring inventors fall prey to consulting companies that charge large fees but don't make a profit for anyone other than themselves. In 1999 Congress took action against invention promotion firms by introducing the American Inventor's Protection Act. Under this law, invention marketing firms must disclose the following.

1. The total number of inventions they've evaluated and the number that have received either positive or negative evaluations;

2. The total number of customers who have hired them in the past five years;

3. The total number of customers known to have received, as a direct result of the particular invention promotion firm's efforts, an amount of money in excess of the amount paid by the customer to that firm;

4. The number of customers who have received a royalty-paying license agreement for their inventions as a result of the efforts of the firm; and

5. The names of all previous invention promotion entities with which the present invention promotion firm has been affiliated within the past ten years.

The American Inventor's Protection Act also permits you to recover for injuries, costs, and legal fees if you can prove a promoter has made any false or fraudulent statements or omissions of any material facts to you in connection with your contract.

---

but bringing your invention out into the sunlight of the marketplace in the early stages is the only way to determine its value in its present state.

Good Ideas versus Marketable Ones

If you show your invention to friends and family, they'll probably tell you it's a good idea. And it probably *is* a "good" idea. But the question you should be asking is: Is it a *marketable* idea? Can you sell enough of the product at a reasonable margin in order to make a profit and justify the investment of time and money? If your goal is to profit from your product, you should develop only marketable ideas.

Do not think that being a successful inventor means you have to put in countless hours of sweat equity and drain your savings being loyal to your idea. You should persist only when you are sure that a financially rewarding market exists, and that your only major issue will be one of *gaining access* to that defined market.

All too often, inventors find that they are the only person invested in or excited about the idea when they attempt to take an invention to market. It may work perfectly, but no one wants it (or at least not at the price at which it can be produced). The problem is not access to the market; the problem is that the market demand does not exist in the quantity necessary to justify the investment.

This confusion leaves many inventors ripe for scams. One major "rip-off" to keep an eye on is the invention promotion companies that tell the inventor he or she has a great idea and promise to land him or her a lucrative licensing contract (see sidebar, previous page).

Ask the Right Questions

So if you can't ask your friends and family for an honest opinion, and you are hesitant to give your money to an invention marketing company, how can you figure out whether to pursue your idea? The answer is with you. As an individual inventor, you must ask yourself the hard questions and do some work to get objective market data before deciding whether to invest time or money in a product.

In fact, you should spend as little time and money as possible developing the actual invention until you have thorough and decisive answers to the questions posed in this book. To start with, you need to know:

WHO WILL BUY THIS PRODUCT? If you are marketing a new technology, how is the industry already functioning without your technology? Why will your product sell, and will the increase in benefits or productivity justify the expense for a consumer who switches to your product?

You must get to know your consumers, delve into their psyches, identify with them, and understand what will lure them away from potentially competing products and alternative solutions. You need to be able to accu-rately define who is the end user, who makes the purchasing decisions, and who actually purchases it. Those things may be done by one person or three. For example, you may have a product that is

---

Whether you plan to license your product or manufacture it yourself, you must get a handle on what you can sell it for and what it will cost to produce.

---

intended to be used by 28- to 34-year-old, college educated males, living in the Southeast, with a household income of $75,000 to $100,000 per year. It's easy to quantify the market size by pulling census data, but what if the actual purchase of the product is done by a spouse? How does this change your model? Understanding the differences between the three variables (end user, influencer, and purchaser) helps determine the market, the market size, and ultimately the marketing strategy.

HOW MUCH WILL USERS PAY FOR YOUR PRODUCT? Whether you plan to license your product or manufacture it yourself, you must get a handle on what you can sell it for and what it will cost to produce. As we've discussed, making an initial prototype is a good way to gauge this. Find out how much competing products cost and what is unique about your product to justify its price. If you know who your customer is, you can get market feedback on price fairly easily. As for what it costs to produce, this is something that you can tackle in a few ways. First, what are similar products (made of the same materials) selling for? You can assume that these products cost approximately 25 percent of the retail price to manufacture. You can also use your prototype to get rough pricing based on the materials used to build it, or to get a quote from a factory that might manufacture it.

The phenomenon of the Pet Rock proved that packaging is everything.

Unfortunately, you never really know *all* the costs associated with a product until you start to produce it. As a product moves through the design stage, across the factory floor, and on to the warehouse before it arrives at the point of retail purchase (this is called the "supply chain"), plenty of costs can accumulate. It's important to identify each link in your supply chain, because each link adds costs. Raw materials, labor, equipment, clerical and other kinds of services, transportation, warehouse space, tooling, manufacturing space, shelf space, packaging, advertising, and customer service are all links. A typical rule of thumb for consumer products is that the overall manufacturing cost for an item should be about 25 percent of the final retail cost of the item, assuming that you sell your product through traditional retail outlets. For example, a product that sells for $40 at retail should cost no more than $10 to manufacture. This allows you to make roughly a 50%

margin while allowing the retailer to do the same.

Once you determine the cost of making the product, you'll know roughly how much you must charge the user. If the consumer perceives the cost of solving their problem as relatively minor, will they pay $40, or more, or less, to solve it? Can you quantify how much time or money your user will save by using your product? Is this time or money important to the user, or could they buy another product that will save them more time and more money? If your product is a novel kitchen item, will users pay its price without hesitating even though it's not a necessity? For example, if you are selling a specific cutting utensil that costs almost as much as a food processor, wouldn't the consumer rather have the multifunctional food processor than a product that performs just one function? On the other hand, if your specific cutting utensil is within the price range of other utensils, and it offers superior features that solve a problem or increase comfort, you may have a competitive product.

The Myth of the Pet Rock

When faced with the tough question "Who will buy your product?" more than one flustered inventor has responded, "Well, look at the Pet Rock. There's a market for everything."

For those of you who missed the fad or are too young to remember the 1970s, the Pet Rock was a rock packaged in a little box that looked like a pet carrying case, complete with "breathing holes." Printing on the box instructed the "pet owner" how to care for the "genuine pedigreed" Pet Rock, and so on. In 1976 it sold 1.5 million units in six months at retail prices ranging from about $7 to $10, with production costs of less than $1 each. The Pet Rock was a small item that fit neatly on store shelves. Consumer tastes were shifting toward novelty items at the time, so an amusing product could be delivered to people at a price they would pay in order to share a laugh—thus the rock filled a defined niche. To give you a sense of the marketplace at the time, shelves were also stocked with lava lamps, posters, and strobe lights.

The phenomenon of the Pet Rock is not evidence that a gullible market will pay a price for a solution to a trivial problem, or that the public is enthralled by the sheer novelty of an invention. In fact, we worry about inventors who cite the Pet Rock in support of their own marketing strategy and argue that "people will buy anything": these inventors are falling into the type of thinking that makes them prey to companies and consultants that offer to bring products to market without asking the inventor to deal with the tough questions. That said, the Pet Rock worked because its inventors put in the effort, did do the research, and *were* able to answer the tough questions—the timing was right, they saw the market, defined it, and priced the product accordingly.

How to Do *Real* Market Research on a Shoestring Budget

Whether your invention is a humor product, a biomedical device, a workshop gadget, a cosmetic, or a software item, you must be able to deliver it at a price people will pay. Large companies spend millions of dollars on surveys, focus groups, and anthropometric research every year. You, as an individual inventor, are likely not to have that kind of money (and you know that expensive surveys and focus groups are not always right in predicting how actual users will ultimately react). Still, there are plenty of realistic ways for you to get reliable data and make the objective assessment you need for your product.

Taking a Simple (but Objective) Survey

We really believe in the value of learning and practicing simple consumer survey techniques—it's not just busywork, we promise. On the contrary, surveys can be uncomplicated and quite effective if they are taken seriously. It's important that you not skip the process of surveying prospective users and not go to market based on your own gut reactions. If you have done a good job of defining who your customer is (at a minimum, age, income, education, geographic location) you should be able to find those customers and assess their interest in your product. Finding them can be as easy as a trip to the mall, the park, or church. Once you know who your customers are, finding them is critical to getting valid data.

Once you get to your potential customers, please be respectful of their time. It helps to explain that you are an inventor working on a new product idea and would value their feedback. Keep the interview under twenty minutes and offer a token reward, such as a gift card or free product, for their time. You will find that these little differences make your prospects more receptive and encourage them to offer honest feedback.

When you have your group, you can ask the participants if they would pay for your product and, if so, how much they would pay. This is a good start. But if you stop there, you'll get only a fraction of the valuable information you stand to gain from them. There are innumerable ways to conduct a survey, and surveys can be as unique as your product itself. The following questions are targeted to elicit some core information you'll need:

How do you currently solve the problem of _____ ?

_____

_____

_____

How much do you pay for that solution?

_____

What do you like about the solution?

_____

_____

_____

_____

What don't you like about it?

_____

_____

_____

_____

If a new product existed that solved the problem in a better way, how likely would you be to try it?

_____

_____

_____

Would you pay more for a product that solved the problem in a better way?

_____

How much more would you pay for it?

_____

Where would you expect to purchase such a product?

_____

_____

_____

How many times a year would you expect to buy it?

_____

You can survey people orally, but handing them a questionnaire to fill out gives the survey an aura of objectivity and increases their willingness to give you honest feedback. The

FROM THE LAWYER
PROTECTING YOURSELF

A nondisclosure agreement (NDA) is an agreement with a company or other party that they will keep confidential the information you disclose to them. As a practical matter, however, NDAs do have limitations. An NDA may be difficult to enforce and is a less secure form of protection than a patent, which documents the full scope of your invention and the date you filed your patent protection.

point of a survey, for an independent inventor, is to get fresh perspectives and hear positive and negative things about your product before you attempt to manufacture and commercialize your invention. Surveys can also help you refine your marketing pitch. We have yet to find an inventor who conducted a survey who was not surprised and enlightened by at least some of the feedback he or she got.

A survey is valuable only if the people you speak with are your potential customers. Ideally you want to be able to survey 200 to 300 people, but even 20 to 30 objective respondents will help you avoid mistakes if they are accurately representative of your customer profile. Select people who you know can be impartial and, of course, avoid steering them toward the "answers."

It's also very important to set the right tone for your survey. People who have agreed to participate will want to please you or, at a minimum, not hurt your feelings if they know you are the inventor or know you in some other way (if they're your in-laws, for instance, or your book club members, your softball buddies, or colleagues). It is important to stress that there are no right or wrong answers and that you are looking for help to identify weaknesses and improve your product.

Keep the survey simple. Some people don't like to write a lot, but you still want their input. Try not to include more questions than you need, and use check boxes, numeric scales, and other tools that can elicit information from the participant simply and quickly. If you know some or all of the competing products, list them for consumers to circle rather than write out. Save the narrative questions for issues like identifying unknown weaknesses or things that stand out. For example, rather than asking what type of laundry detergent the person uses, ask the person to circle which type of laundry detergent is used in the household: A) Tide, B) All, C) Cheer, D) Other (please fill in).

It is not necessary to disclose your invention as part of the market research. You can simply discuss the current problem and how consumers address it. If you do plan to reveal specific details or drawings of your invention, however, have each participant sign a nondisclosure agreement (NDA, like the one in

Appendix A) before they start the survey—even if your product is "patent pending." If you have a product that is not yet patented but is potentially patentable, nondisclosure agreements are an absolute necessity.

If you can, it's a good idea to protect your product with a provisional patent application before showing it to people in a survey. The provisional patent application gives you stronger legal protection than an NDA does. A provisional patent application is a type of "temporary" *application* (the word "provisional" refers to the application itself) that gives you patent pending status, but it allows you to make modifications during the first year after the public disclosure of your product (telling someone about your product in any way—including surveys—even if you don't sell it). See Chapter 4 for more on provisional patent applications.

Finding Venues to Test-Market Your Product

For an individual inventor, test marketing is one way to take as little financial risk as possible until you know whether the market response is what you had predicted. Test marketing gives you an opportunity to make revisions and modifications before larger production runs. As an individual inventor, you can even test-market before you start producing the product.

ONLINE SALES Selling a product online is an excellent way to test a new product.

> # If you have a product that is not yet patented but is potentially patentable, nondisclosure agreements are an absolute necessity.

Millions of people cruise the Internet looking for solutions to their problems. Discussing your product in an online chat room or blog (once it's appropriately protected with a patent application, of course) serves to introduce your product to a very targeted market. In addition to online marketing, many large companies advertise products on television and offer limited sales on their website, sometimes before they even produce the product. If they don't have the product on hand, they offer a projected shipping date or sometimes indicate that the product is sold out. The companies use this Web marketing technique to get valuable data about who will buy the product if they offer it. Just think of all the information you enter when you place an order on the Internet! Sometimes, if the companies don't get the initial market response they want (actual orders), they refine the product, adjust pricing, or in some cases, cancel the launch of the product altogether.

TRADE, HOBBY, AND ELECTRONICS CONVENTIONS What better way to meet prospective customers with shared needs and interests than by going to a

EdisonNation.com:
Fast-Track Your Ideas to Market Online

Edison Nation (www.EdisonNation.com) is an online community for inventors and idea people from the producers of the Emmy award–winning PBS series *Everyday Edisons*. The site provides a platform for the world's leading retailers and manufacturers to host searches for innovative new ideas—allowing inventors to fast-track their ideas to market quickly and inexpensively. Here's how you might use Edison Nation to advance your ideas:

- **Get Your Ideas in Front of the Big Guys:** The most revolutionary concept introduced by Edison Nation is a "Live Product Search." Rather than forcing inventors to take educated guesses at needs in the marketplace, major retailers and manufacturers tell the community what they're looking for. Sometimes it is very broad: "We're looking for a new sporting good." Other times, it is very defined: "We're looking for new protective body armor for high school football players." New searches are launched all the time and at any given moment a Live Product Search may be launched that covers *your* product or area of expertise!

- **Learn From Invention Experts with Full-Screen Streaming Videos:** Watch *Everyday Edisons* episodes online and behind-the-scenes interviews with members of the show's product development team and inspirational interviews with innovation experts.

- **Browse Electronic Archives of *Inventors Digest* Magazine:** Access years of informative articles, tips, and tricks when you browse, read, or print electronic archives of *Inventors Digest*, the voice of the independent inventor community.

- **Find People and Groups Who Can Help:** The online social networking components of Edison Nation help inventors connect to each other one-on-one and in idea-centric groups. Inventing is a group effort—Edison Nation can help you build your team (and your success).

- **Get Fast and Free Answers from a Forum:** Most independent inventors want to help each other. For that reason, the discussion forums are an extremely lively and popular part of

gathering where a lot of them will be present? Even if you cannot afford to be an exhibitor (booth fees can run from $1,000 to $10,000 or more), taking your product around to show to other attendees and visitors can be invaluable. We know of one young inventor who took a patent-pending craft-related game to a hobby show and walked the floor for two days asking people if they might be interested in buying it when it became available. If they answered yes, she had them sign their name and contact information on a legal pad. Her

Edison Nation. The forums contain thousands of topics at any moment, teeming with valuable advice from successful inventors—including many industry experts. Discussions are updated in nanoseconds, and requests for help are met and validated by thousands of inventors typing on topics of common interest 24/7. Current forum topics cover patents, prototyping, branding, packaging, selling, market research, and so on.

The world's leading retailers and manufacturers are challenging inventors on Edison Nation with new Live Product Searches all the time. And Edison Nation members are not only seeing their ideas hit store shelves in record time, they're getting paid for them.

The quickest way to get a great idea in front of a major retailer or manufacturer is to submit it to a Live Product Search on Edison Nation. Innovation-hungry corporations review ideas submitted to them through Edison Nation, and license the best of the best.

Licensing an idea (as we explain in Chapter 6) is a great way to profit from your concept when you don't have the capital to develop it. Edison Nation pays a generous royalty and upfront advance if your product is selected. (See the website for current details on searches and submission rules.)

Participating in an Edison Nation Live Product Search requires that you submit your idea online, with a commitment that the idea will be kept confidential by the Edison Nation team and the participating corporations. It's not necessary to have a patent, prototype, or formal drawings. The Edison Nation team has a well-earned reputation of respecting inventors and honoring its confidentiality commitments.

In the event your idea doesn't fit within the bounds of the current Live Product Searches or you just feel you're too far along to disclose details without a traditional face-to-face meeting, you'll find help in the forums and other educational features offered on the site for making connections to pitch those great ideas corporations didn't even know they were looking for.

feet hurt at the end of the show, but her legal pad was overflowing with prospective purchasers and how to contact them. This list proved invaluable in convincing one company to underwrite her manufacturing costs and convincing another to give her royalties under a licensing agreement to distribute the product.

SPECIAL EVENTS Fund-raisers, silent auctions, and other community events can be great ways to build goodwill and gauge market response to your

FROM THE ENTREPRENEUR
CONVINCING YOURSELF

Getting feedback early on doesn't just give you a notion of who your potential buyers are; it makes your idea more tangible and helps turn it into something that can be shared with more people, which is the first step in selling the idea as a product. It helps convince consumers that your product is valuable and, most important, it serves to confirm the same to *you*.

product. Donating or offering the completed samples of your product at a reduced price may be worth the return in advertising and exposure. One inventor we know had a patented educational puzzle toy with intriguing packaging and a companion website for kids. The inventor made his product available for school fund-raisers at a cost below retail, but required the school to sell it at his projected retail price. He tracked his sales carefully and had the recorded data in hand when he recently encountered a vice president of sales whose child was particularly enamored with his invention. He convinced the VP to set up a meeting.

Another inventor we know donated several of her products to a silent auction and found that attendees weren't willing to bid her "retail" price, even with part of the proceeds going to charity. Since those people were representative of the inventor's target market, she came away

knowing she had to concentrate on bringing the cost down.

SMALL, INDEPENDENT STORES As we will discuss in Chapter 5, large retailers and corporations can be difficult to penetrate until you have research and proven sales behind you. But there are plenty of small, independent stores in your own community where you can talk to an owner to get your product onto the shelves. You can negotiate a price, commissions, and how and when you will take it back if it doesn't sell. One inventor we know is a nurse's aide who started selling limited quantities of her product to hospital gift shops; she personally restocked them as needed. She now has a credible sales history to take her patented concept to major greeting-card and novelty retailers, not to mention valuable data from talking directly to the shop managers.

Test marketing may seem like a big leap, but whether you're planning to manufacture a product or going for a license, the sooner you talk to real customers, the better. Getting "preorders," even small ones, from actual purchasers (as opposed to survey subjects) will add to your credibility. Your product's prospects for success will always be improved by moving it forward into the light of day and letting the market speak to you as early as possible.

CHAPTER 3

Prototyping, Manufacturing & Distributing

A Crash Course

Earl Tupper was born in 1907 to a New Hampshire farm family that took in laundry and ran a boarding house to make ends meet. Earl's father, Earnest, was a "barn inventor" who scraped together enough money from the family's tight budget to get a patent for his chicken-cleaning device. Though the chicken-cleaning frame was never mass-produced, it fostered an inventive mindset in his young son, who would grow up to invent Tupperware and create a radical distribution model for it.

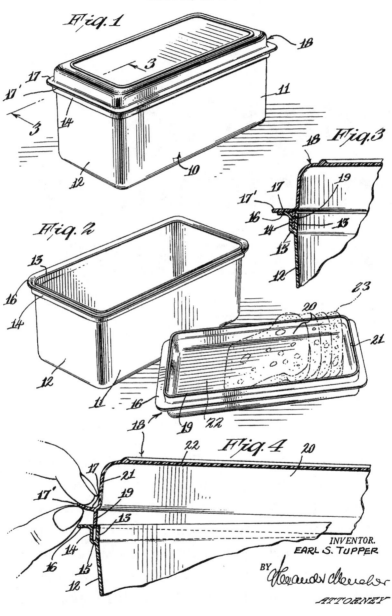

Nov. 30, 1954   E. S. TUPPER   2,695,645

BREAD SERVER OR ANALOGOUS SEAL TIGHT CONTAINER

Filed May 8, 1950

Earl Tupper invented an air-tight container that kept food fresh and prevented leakage.

At the age of ten, when he began selling his family's produce door-to-door by horse and buggy, Earl discovered that consumers would buy more of a product if he took it to them. Throughout high school and the Great Depression, Earl remained optimistic and entrepreneurial and kept written scientific notebooks about the inventions in his head. He was befriended by Bernard Doyle, the inventor of a plastic called Viscoloid, who helped him get a job at DuPont in 1937. Earl stayed only a year at DuPont, learning about research and development, design, and manufacturing, before starting his own company, which specialized in plastic products based on the ideas filed away in those notebooks and in his head. Quitting a well-paying job during the Depression was risky, especially because the availability of raw materials used to make products of plastic was limited.

Regardless, the Earl S. Tupper Company thrived, advertising its plastics design and engineering capabilities (mostly to industry). Many of the company's early contracts were subcontracts from Tupper's former employer, DuPont. One, for instance, was a project to turn cheap pieces of brittle, greasy, smelly polyethylene slag into gas masks and lightweight containers for the army. It was during this time that Tupper invented and patented his famous (grease- and odor-free) spill-proof containers with their airtight lids. (His inspiration? The lid of a paint can!)

The Tupper Strategy for Locking Down Profits

T upper proved ingenious at using low-cost materials to manufacture his products and securing wide-reaching and innovative distribution channels to help sell them. For example, he got the huge Tek toothbrush company to offer a Tupperware tumbler with purchase of a toothbrush. He convinced large cigarette companies to offer Tupperware cigarette cases imprinted with the companies' logos. However, when his products first debuted in department stores, they gathered more dust than interest—in fact, they were a flop. Consumers didn't see the benefits of the new airtight lid without hearing an explanation, as well as seeing a demonstration.

Realizing this, Tupper revamped his distribution strategy. He noted that

In-home Tupperware parties revolutionized selling and defined an era.

two distributors, Thomas Damigella in Massachusetts and Brownie Wise in Florida, were moving a small but steady inventory of his products through in-home demonstrations. Tupper paid attention. In 1948 Tupper convened a meeting with Damigella, Wise, and several other small local distributors at a Sheraton Hotel in Worcester, Massachusetts, to tell them his idea for a new Tupperware sales campaign. It was at this meeting that Tupper launched the home sales-party plan that became and remains, decades later, a profitable outlet for Tupperware. In 1958 Tupper sold his operations to Justin Dart of Rexall Drug Company for $16 million. Tupperware currently employs more than 11,000 people and has gross revenues of nearly $2 billion, and the home party system (often called the Tupperware party), supplemented by its catalogs, continues to thrive.

Tupper understood that the key to profit was more than just finding a target market. He had to fashion his product using available, cost-effective materials and manufacturing methods. And, equally important, he had to get his product into the outlets where the target market would find and use the product.

Manufacturing and Distribution

anufacturing doesn't need to mean setting up your own factory and going into debt to buy equipment. As an individual inventor, you can do the very same thing large corporations do: outsource to an existing manufacturer. In fact, you will probably outsource to several companies. For example, you may hire separate companies to produce your parts, assemble them, warehouse your product, and fill orders. To determine the costs of producing your product, you must research all phases of your manufacturing and distribution process, including the following:

■ **Determine the price your consumers will pay for your product.** Even consumers who strongly want your product generally have a limit to what they will pay; you must be able to deliver it at an acceptable price point.

■ **Figure out if you can manufacture your invention for less than 25 percent of what consumers are willing to pay.** Our experience shows that you must be able to produce your invention within this constraint to be able to sell it profitably in major retail outlets. If you are selling directly to a consumer, you may be able to justify higher production costs, since you don't have to pay a middleman.

■ **Design your invention and make one or more prototypes.** You will need to make a working version to get an idea of production costs, and ultimately have to produce a presentable packaged product to sell to consumers, distributors, and retail outlets.

■ **Find manufacturers capable of properly assembling your invention.** Even if it's simple, you will generally outsource the manufacturing to someone else rather than investing in equipment, staff, and personnel.

■ **Identify your process of fulfilling orders and shipping your invention.** If you are soliciting orders from major retailers, many of their questions and concerns will be directed at your ability to fulfill orders and ship on time. Even if you sell your product primarily through small retailers or online sales, you will need a reliable method for order fulfillment.

■ **Secure as many distribution channels as possible.** A distribution channel is basically any venue that gets your invention into the hands of paying consumers, including, but not limited to, mass merchants, "big-box" stores, specialty retailers, small mom-and-pop stores, websites, shopping channels, and catalogs. The more distribution outlets there are that are appropriate for your product, the more product you'll be able to move.

■ **Expand and evaluate your distribution channels to maximize sales.** Many inventors introduce their product to the market through small distribution channels such as local stores, catalogs, and websites. As a product develops, proven sales establish "market traction," which indicates that a company has achieved a consistent level of sales

within a market segment (which helps in predicting future success of the product in other outlets).

Does Manufacturing Make Sense for You?

Why not bypass the whole supply chain and skip to the front of the line by licensing your product (and collecting a tidy royalty check)? Why should you consider distributing a product yourself when other companies already have distribution channels in place? Distributing on your own involves far more risk and work than licensing does. However, it is also the path to far more profit. Having a patent pending product or even a patented one to sell to another company is never as financially valuable as having a proven product that can be successfully manufactured and sold for a profit.

Running the Numbers: Generating Sales Revenue versus Receiving a Royalty

Deciding whether to license your product instead of selling it comes down to deciding between risk and reward. In the dynamics of the marketplace, it will always be the person who commercializes the product, rather than the inventor who licenses it, who stands to benefit most. Conversely, commercializing comes with the greatest risk. Typically, inventors realize a 20 to 50 percent gross margin on a product they manufacture themselves. If they decide to license their patent instead of making the item themselves, they may get a 2 to 5 percent royalty based on net sales. Let's consider an example that illustrates the profit potential of manufacturing compared to licensing. Suppose you have a patented widget. The widget costs $4.00 to manufacture and sells to users at $16.00; you stand to make $4.00 to $12.00 on every unit sold, depending on whether you sell it directly to the consumer or through a retailer. Using a typical retail relationship as an example, you sell the product to the retailer for $8.00, thus creating a 50 percent gross margin. The retailer in turn sells the product for $16.00, also making a 50 percent gross margin. On the other hand, if you license it to a company that is already in the industry, and that company pays you a standard royalty rate of 2 to 5 percent, you may make only 16 to 40 cents per unit. The royalty is paid on the wholesale price of the item ($8.00), not the retail price.

Suppose you are fortunate enough to have your product picked up by just one major retailer that orders 10,000 units of the widget directly from you. As a manufacturer, you stand to make a $40,000 gross profit. On the other hand, if you license your patent right to the widget to a company at a 5 percent royalty rate, your gain is limited to about $4,000. The difference here, roughly $36,000, is an example of risk and reward. To obtain the $4,000 royalty, you have

not assumed any additional risk to get your product to market. The licensee, who paid you the $4,000, assumed all the necessary costs of producing and distributing the product. Their gross profit is far greater than your royalty; however, they incur significant expenses to operate a business and distribute your product.

Producing a product yourself can be more profitable than licensing (a $40,000 gross profit is much greater than a $4,000 royalty); however, as the manufacturer, you now bear the costs of running a business devoted to producing and/or marketing, selling, distributing, and servicing the product. Your net profit—the amount of profit left over after you subtract your selling, general, and administrative expenses (SG&A) from your gross margin—can quickly erode based on your SG&A costs of doing business.

Leverage Your Competitive Edge as the Inventor

The law allows you, as the inventor, protection when bringing your product to market. Both U.S. and foreign patent laws also give you a huge running start over all your competitors in the marketplace. Remember, if you hold a U.S. patent, you get to exclude others from making, selling, or using your invention for a twenty-year period. By limiting others, a patent gives you these rights exclusively unless you decide to license it to someone else. (Turn to Chapter 4 for more about how the patent system keeps the competition from breathing down your neck.)

How Can a Truly Great Product Not Have Good Licensing Prospects?

There are many profitable products that can and should be brought to market even if the inventor is not successful in finding a company to license his patent. Not every fledgling product, even if it is superior to the competition, will find a home with an established manufacturer in an existing product line. Sometimes the product is too innovative, "ahead of its time," or deemed disruptive to the current marketing strategies of prospective licensing companies.

It goes without saying that the companies you approach to license will be those that offer your invention the best fit in their product line. You must remember, though, that you are looking to bring a new product into their existing market space, and that can work against you in procuring a licensing agreement. The companies you approach have a lot invested in marketing their current product lines and ensuring that they are recognized and accepted by the consumer. Why should they erode profit margins they have worked hard to build up by offering a product that may "cannibalize" their existing market share? Even if your invention is a superior product, introducing it may have

the net result of splitting up their existing market.

Of course, viewed from a licensee's perspective (which is the perspective that counts during a pitch), a product that decreases profits would never be a good business decision. Companies are, understandably, in business to make a profit, and no business can stay around for very long if it makes revenue-reducing decisions. At the time you approach a company, it may or may not be in the company's economic interest to give consumers a better choice if it has already invested in and developed a marketing strategy to bring a competing product of its own to market. (You may have to wait it out to see how successful the competing product is, and possibly approach the company again when the timing is better.) Thus, your invention may not be attractive to companies as a licensing prospect for reasons having nothing to do with the product's actual profit potential. In fact, it may be in their economic interest to hope your invention never sees the light of day. The sheer economics of innovation may push you into the manufacturing arena.

Deciding to "Disrupt" the Existing Market on Your Own

Companies may feel no particular urgency to develop your invention to stay competitive if they don't think you or anyone else will successfully get it to market. However, beginning the

(continued on page 56)

CASE STUDY
Storming a Saturated Market

It is possible to enter a crowded field (such as strollers and other baby gear) by pitching your idea carefully and in such a way that companies see your invention as being in a class of its own rather than something that stands to erode their existing market. This is exactly what best friends, do-it-all moms, and inventors Lynn McIntyre and Karen Madigan did when they pitched their patent pending stroller.

Ever since they bonded as the "only altos in junior high choir," Lynn and Karen have looked to each other for support and advice, even through college, marriage, kids, and careers. So when Karen couldn't figure out how to keep her youngest son content in a stroller, she immediately turned to Lynn. After brainstorming dozens of concepts, the dynamic duo had a eureka moment: Why not combine an activity saucer with a stroller so a child could stand up, spin, and explore *or* sit and rest? Karen and Lynn invested thousands of dollars and countless hours over several years just to have their idea designed before they made their way to the *Everyday Edisons* team.

Today their stroller has been branded Kineta, and it gives children the freedom to wiggle, spin, and bounce as long as their boundless energy can sustain them. A unique rotating arm affords the child a 360-degree view and lets him move about to his heart's delight. With a revolving seat that locks into place, stroller-bound children can see the park from a whole new perspective—keeping kids content and parents from pulling their hair out! With this distinct consumer market, Lynn and Karen anticipate that their device will be seen by prospective retailers and licensors as a device in its own class, rather than just another competing stroller. Bringing a product as complicated as a stroller to market is a challenge. In addition to complex mechanical and manufacturing considerations, there are numerous safety concerns and issues related to a product that carries such precious cargo. There is also a steep learning curve for anyone who enters a market with a product that is disruptive to a category.

Sometimes success can take longer to realize than you expect, and there are no guarantees that your product will ever be accepted. On the other hand, you can't win if you don't play. Perseverance is critical, and so is aligning the right resources, contacts, and relationships.

Like a mini playground in a stroller, Kineta allows children movement without straying from their parents.

US 20070246915A1

(19) **United States**

(12) **Patent Application Publication** (10) Pub. No.: **US 2007/0246915 A1**

Madigan et al. (43) Pub. Date: **Oct. 25, 2007**

(54) **STANDING BABY STROLLER**

(75) Inventors: **Karen Jacy Madigan**, Huntersville, NC (US); **Lynn Furton McIntyre**, Huntersville, NC (US); **Daniel Lee Bizzell**, Davidson, NC (US); **Ian Douglas Kovacevich**, Charlotte, NC (US); **Kevin James Dahlquist**, Charlotte, NC (US)

Correspondence Address:
TILLMAN WRIGHT, PLLC
P.O. BOX 471581
CHARLOTTE, NC 28247 (US)

(73) Assignee: **STROLLER TECHNOLOGIES, LLC**, Charlotte, NC (US)

(21) Appl. No.: **11/767,500**

(22) Filed: **Jun. 23, 2007**

Related U.S. Application Data

(63) Continuation of application No. 11/175,602, filed on Jul. 5, 2005, now Pat. No. 7,234,722.

(60) Provisional application No. 60/584,991, filed on Jul. 2, 2004.

Publication Classification

(51) Int. Cl.
B62B 7/00 (2006.01)
(52) U.S. Cl. 280/642

(57) **ABSTRACT**

A stroller for transporting a child includes a wheeled frame including at least two base rails interconnected by a cylindrical hub, a spine, and a bracing linkage. A basket attaches to the spine, confining the child and supporting a seat ring. The seat ring includes upper and lower connected concentric arcs having different radii and supports a seat back on its lower arc. The seat back reclines, and the seat ring rotates from center. A seat base is supported on the frame and includes a platform movable between an extended position disposed inside the seat ring and a retracted position withdrawn from the seat ring. The stroller permits the child to be safely and comfortably transported in a standing, seated, or reclined position.

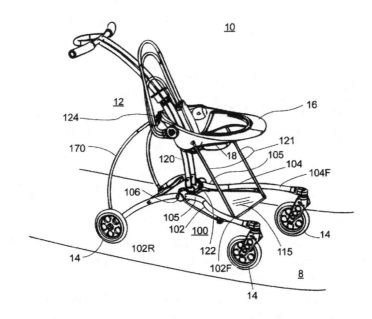

Karen Madigan and Lynn McIntyre's standing baby stroller patent.

(continued from page 53)

manufacturing yourself and acquainting consumers with your alternative product may be just the push companies need to understand that you will become their competitor. Proven sales demonstrate that you have a product that is not going away, and that you have the long-term capability to disrupt their market share if they do not jump aboard. In short, manufacturing can be a path to a lucrative licensing agreement.

It's important to understand that companies have a relatively conservative mindset in licensing products. They want assured gains for bringing on a new product, and they want assurance that it won't disrupt their existing profits and market shares. But true innovation is almost always disruptive, so manufacturers of existing products may initially view it as competition to be squelched rather than fostered.

Taking the Terror Out of Manufacturing on Your Own

Licensing is initially appealing to most new inventors because the prospect of producing and selling a product on their own seems terrifying. Besides the costs, manufacturing involves acquiring a whole new skill set: navigating unfamiliar contractual arrangements and learning about methods of manufacture. An experienced manufacturer may have negotiated similar terms dozens of times and is

totally familiar with what's considered standard, but the terms, as well as the technologies, will likely be entirely new to you and utterly confusing.

Surfing to Success

The good news is that you can research and learn everything you need to know about manufacturing your invention. Since you had the wherewithal to invent something marketable, you already have the chops to market it successfully. There is no one more qualified or passionate about the product than you, no one more familiar with how it works and why it is necessary. However, you must be willing to work hard to advance yourself and your invention by research and refinement.

You must do your research before you sign any checks so you don't run the risk of others' taking advantage of you. Some books and advisers may tell you that you can make and sell an invention without using a computer. On the contrary, being comfortable with computers is an important part of the entrepreneurial process, particularly when you're researching who can produce your product, how much it will cost, and whether estimates you get are in the ballpark. You'll need to spend a lot of time e-mailing, surfing the Net, following up on information, and cross-referencing information from other sources. If you do this diligently, though, you'll find access to the data you need, even if it's not always the answer you want. You'll have the information at your disposal

CASE STUDY

Overcoming a Complacent Market

Vacuum cleaners are a good example of a complacent industry with established technology that was ripe for disruption. In 1979 an inventor named James Dyson became frustrated when his top-of-the-line vacuum cleaner began clogging and losing suction because, like other vacuums, it relied on a bag and filter to store removed dirt. Both items clog, obstructing airflow and diminishing the performance of the vacuum. Dyson was determined to come up with better technology. Fifteen years and, according to his own estimates, 5,127 prototypes later, Dyson developed the first vacuum cleaner that doesn't lose suction. He knew that the small pores in vacuum bags and filters clog with the fine particles of dust; he decided there had to be a better system and invented it. Dyson's patented cyclone technology creates a centrifugal force to remove dust and dirt, along with providing an unobstructed clog-proof airflow (see next page).

Although it was easy for Dyson to prove his vacuum worked better and was more cost-effective than others', he spent two fruitless years attempting to license it to large U.S. and British companies, including Hoover, Black and Decker, and Electrolux. All of them turned him down, hesitant to undermine the profitability of $500 million in vacuum bag sales and to erode the profitability of products they'd already designed and developed. Dyson decided to bring his bagless DC01 to market himself, overcoming countless financial obstacles and investing his personal resources at a time when others would be feathering the nest for their retirement. By 1995 it became the bestselling vacuum cleaner in the U.K.

In the end, the superiority of Dyson's product prevailed. His first upright vacuum cleaner, the DC07, appeared on the U.S. market in August 2002. In less than one year, the public's awareness skyrocketed. Practically all retail stores that sell vacuums in the U.S. carry the product. Dyson is the bestselling vacuum cleaner in Western Europe, Australia, and New Zealand.

Dyson publicly advises inventors with improved technologies to have high expectations for their own success. He advises them to "break down an imposing task into smaller, manageable ones, believing that you are able to achieve your goals, whatever they may be. Be dogged and determined—and don't be afraid to be different."

Convinced there were improvements to be made, James Dyson tackled a complacent vacuum cleaner market with a superior product.

United States Patent [19]

Dyson

[54] **UPRIGHT VACUUM CLEANING APPLIANCE**

[75] Inventor: **James Dyson,** Bathford, England

[73] Assignee: **Prototypes, Ltd.,** Bath, England

[21] Appl. No.: **655,148**

[22] Filed: **Sep. 28, 1984**

Related U.S. Application Data

[63] Continuation-in-part of Ser. No. 452,917, Dec. 27, 1982, abandoned, and a continuation-in-part of Ser. No. 627,110, Jul. 2, 1984, abandoned, and a continuation-in-part of Ser. No. 627,292, Jul. 2, 1984, and a continuation-in-part of Ser. No. 628,346, Jul. 6, 1984.

[51] Int. Cl.^{4} **A47L 9/16; A47L 5/32**
[52] U.S. Cl. **15/335; 15/339;**
15/352; 15/391
[58] Field of Search 15/352, 350, 347, 339, 15/366, 383, 391, 335

[11] **Patent Number:** **4,571,772**

[45] **Date of Patent:** **Feb. 25, 1986**

[56] **References Cited**

U.S. PATENT DOCUMENTS

1,759,947	5/1930	Lee	
2,184,732	12/1939	Brewer	15/352 X
3,040,362	6/1962	Krammes	15/352 X
3,482,276	12/1969	Fillery	15/391 X
3,634,905	1/1972	Boyd	15/350
3,790,987	2/1974	MacFarland	15/328 X
4,377,882	3/1983	Dyson	

OTHER PUBLICATIONS

Technology–Apr. 25, 1983.

Primary Examiner—Chris K. Moore
Attorney, Agent, or Firm—Ian C. McLeod

[57] **ABSTRACT**

A vacuum cleaning appliance having dual spaced apart air conveying pipes (**13, 14**) supporting a cap (**15**) for a dirt container is described. The pipes are supported on a casing (**11**) mounting a movable cleaning head (**10**). The appliance is also convertible to a tank type cleaner using a handle (**30**) for the appliance.

20 Claims, 11 Drawing Figures

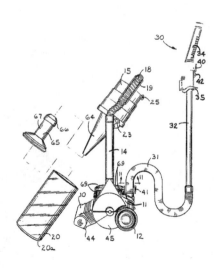

James Dyson's new and improved vacuum technology, patent no. 4,571,772.

to know when to pursue a project and when to move on.

If you do not feel entirely comfortable with Internet research and setting up professional-looking e-mail accounts, help is available to you for free at your local library. It is part of a reference librarian's job to assist you with research and help you refine your computer skills. They will teach you how to use a computer, even if you are logging on for the very first time. They will help you find companies, locate contacts, get manufacturing quotes, learn regulations and standards for your product's industry, and even help you do patent searches. This information could cost thousands (if not tens of thousands) of dollars if you hire someone to do it for you. Our dear librarians are the unsung heroes of inventors everywhere. This chapter will teach you where to start your search and the right questions to ask. You'll have no excuse not to learn how to do research on your own behalf, even if you've never logged on to a computer in your life.

It is up to you to become the foremost expert on manufacturing and marketing your invention. As the initial sole advocate and promoter of your idea, you've already established that your invention is novel and fills an identified need, which means you already have the instincts to direct the research and spot manufacturing and marketing issues as you delve deeper into the process.

Optimistic inventors are the target of scams, excessive markups, bogus mar-keting services, and unfulfilled promises that come at a considerable cost. Doing Internet research is the best way to protect yourself. Inventors are vulnerable mainly if they lack the initiative or ability to research product pricing and market data themselves.

FROM THE ENTREPRENEUR
DO SOME SHOPPING

Though the dot-com decade made a lot of people into billionaires, you'd be mistaken in thinking that inventing is all about staring at computer screens—it's about educating yourself about the ever-changing consumer market. Knowing the market requires research—and not just among Web pages. Visiting trade shows, scanning store shelves for current items within your product category, reading industry publications, and familiarizing yourself with the state of the consumer market (i.e., following the stocks, keeping up with news on the economy) are all paramount in creating a successful product.

Structuring Your Research

Even if you are adamant about wanting to license your patent instead of manufacturing your product, you must know how much it will cost someone else to produce your invention. You can't pitch a product unless you are able to convey its profitability, and you can't do that unless you know how much it will cost the licensee to manufacture it. It is also important to determine what a fair and reasonable royalty would be for your invention. While the industry average is 2 to 3 percent, lower and

Costs to Consider

Perhaps the most universal obstacle inventors face in turning a raw idea into a product is coming up with the necessary funds at each stage of development in order to keep the process going. The following is a critical checklist of costs to consider in your budget.

- ❑ **Conceptual design fees:** initial costs of designing your conceptual idea

- ❑ **Early prototype:** expenses to create "proof of concept" models to test the idea

- ❑ **Patent fees:** payments required to protect your idea

- ❑ **Engineering:** fees paid for having the product engineered

- ❑ **Sourcing:** costs to locate a manufacturer (travel) and secure manufacturing (samples, for example)

- ❑ **Final prototype:** costs of ordering or producing the final "looks-like/works-like" prototype

- ❑ **Tooling:** set-up fees and other costs associated with creating the molds or dies to begin production

- ❑ **Initial inventory:** cost of having product available for sale

- ❑ **Shipping of samples:** cost of transporting initial product samples from your manufacturer to you

- ❑ **Quality assurance:** cost of having the product inspected before shipping

- ❑ **Product testing/quality assurance inspection:** fees for outside laboratories or agencies to inspect and test your product

- ❑ **Freight:** expenses of 1) transporting your shipment of product to retailers and 2) delivering it to customers

- ❑ **Duty:** taxes imposed on your shipment if it originates outside the United States

- ❑ **Customs:** fees imposed on your shipment upon entering the United States

- ❑ **Warehousing:** cost of storing your product inventory prior to distribution to your customers

- ❑ **Fulfillment (picking, packing, and shipping):** cost of managing your inventory before it's received by your customer

- ❑ **Credit card processing:** the fee for accepting credit cards as a customer payment option

- ❑ **Customer service (customer support line):** costs of managing the ongoing relationship with your customers

- ❑ **Product liability insurance:** cost to indemnify yourself (or your company) in the event someone is harmed or their property is damaged as a result of or while using your product

- ❑ **Business license(s):** costs for filing for and procuring license

higher amounts could apply based on the anticipated cost to produce, retail price, and volume of sales. Most inventors—even seasoned ones—forget to count up all the costs at the beginning.

You will need to research costs at all phases of the development process. (See the box on the facing page for a checklist of cost considerations.) There are several phases in the conception, design, manufacturing, and distribution process to take into account. Generally, most inventions go through three major phases:

1. Design a product and price prototypes. You must initially design and engineer an invention that works as planned. You must also be able to demonstrate to others how it works by making a sample, or prototype. James Dyson, discussed earlier in this chapter, states that he has made more than 5,000 prototypes, changing his initial working version one feature at a time. Many inventors make several prototypes at various stages of development, with the final prototype being a near finished sample. Start with a proof of concept prototype with parts you obtain and assemble on your own, and then move toward increasingly more polished prototypes as you refine your invention.

2. Get quotes for initial production runs. Whether you are licensing or manufacturing your invention, you need to know how much it will cost to make it. Licensors will want to know this to assess whether your product will be a profit-able addition to their product line. You will certainly need to know how much it costs to get your product on store shelves, so you can determine an appropriate selling price and profit margin.

3. Package your invention. A large percentage of inventions become consumer products sold through traditional retail channels. Some great inventions, however, are sold through more commercial or industrial channels of distribution. These products rarely require extensive or particularly attractive "retail packaging"; they instead take a more utilitarian form of packaging (protecting the contents). For products being sold through traditional channels, packaging accounts for those very important last three seconds and last three feet before a consumer makes a decision.

In doing initial cost projections, individual inventors often fail to adequately research the costs of designing and manufacturing packaging. If the packaging is inadequate or presents a poor appearance, retailers may make a snap judgment that your product is not market ready. As part of this phase of your research, you'll need to figure out how best to display your product (is it better as a hanging item or stacked on a shelf?). You'll also need to consult with someone who has design expertise to determine how your product can be shown to its advantage without taking up too much shelf room.

Although the initial research steps outlined above are the same for both licensing and manufacturing, you do not need to commit to either one at the outset. In fact, you're advised to keep as many options as possible open for as long as possible, until you feel confident in deciding which avenue will be the most profitable and accessible for you.

Keep as many options as possible open for as long as possible, until you feel confident in deciding which avenue will be the most profitable and accessible for you.

Some Websites to Get You Started

There are a number of free online resources to help you find a good prototyping or manufacturing service. We recommend that you become familiar with the following:

■ **The Thomas Register of American Manufacturers:** This free online database is located at www.thomasnet .com. It covers approximately 174,000 companies, both public and private, classified under more than 72,000 product and service classes. You can find more than 650,000 distributors, manufacturers, and service companies within more than 67,000 detailed industrial categories. To use this directory, simply put in a keyword that reflects your product category. For example, in the illustration below, "plastic injection molding toy" was the search term entered.

■ **The Thomas Register of Global Manufacturers:** Another searchable Web directory is www.thomasglobal .com. It includes more than 700,000 industrial suppliers located in major markets worldwide. Like the Thomas Register of American Manufacturers, this database is constantly updated. It currently contains information about 3,000,000 product listings and is organized by 11,000 keywords. Even if you are new to manufacturing, you can use it to e-mail foreign suppliers to request quotes and other documentation.

■ **Alibaba.com:** This is another site for researching international manufacturing options, which you can use to cross-reference information you find on thomasnet.com and thomas global.com (discussed above). The site has millions of listings across more than 5,000 product categories.

■ **MFG.com:** This site is set up as an online marketplace, in which suppliers and manufacturers ("sellers") pay to advertise their services to inventors. It is free for inventors ("buyers") like you to use. The "sellers" pay an annual membership fee to promote their services on the site.

■ **EdisonNation.com:** This is an online community where inventors can get information on local resources from other inventors. For more, see page 44.

Protect Your Confidentiality While Seeking Quotes

In order to communicate effectively with manufacturing sources, you'll need a set of drawings with detailed specifications about how to make and use your invention. Even if you have a patent pending, you want to take reasonable steps of your own to keep this information proprietary and ensure that it is not examined or disclosed without a purpose that is to your benefit and related to what you've asked the company to do for you.

When you are disclosing your idea to engineers, industrial designers, and manufacturers, it is good business practice to ask that they sign a nondisclosure agreement (NDA) first, specifying that they won't use confidential information you bring them for any purpose other than quoting and producing your product. (A brief sample NDA is included in Appendix A). This is a fair request, and most companies are honest. They know that if they earn a reputation for stealing ideas, they won't be around for very long.

The agreement in Appendix A is deliberately simple, so that companies can read it quickly and will be willing to sign it for you on the spot. It is a good starting point for inventors who do not have a lot of leverage in the initial stages of negotiating.

Some large companies may refuse to sign even the simplest bare-bones agreement and may instead ask you to sign theirs. Although this is appropriate if you are approaching a company with a patented idea for licensing, it raises some issues if you are bringing manufacturing specifications to them. These specifications are far more valuable than an undeveloped idea.

If you elect to sign a company's agreement and are taking them your research in any form, make sure that the definition of "confidential information" includes information about your invention that was not known to them previously. Read the agreement and ask questions to make sure that it reflects your expectation that, at a minimum, confidentiality be extended to your pending patents as well as issued ones.

If you feel that you have a good relationship and sufficient leverage with a manufacturing company, you might want to include some of these additional provisions in the agreement:

■ In some situations, the manufacturer may actually help make improvements to your product. Under such circumstances they become a coinventor, so it is important to define who owns the resulting intellectual property. A clause that states the company will assign any intellectual property to you that they develop on your behalf, and will help you secure patents on these processes and modifications made at your expense to improve your product addresses this issue. These innovations can be valuable to you in the future and should remain your property.

■ The company will not use any information obtained from you (whether confidential or not) to compete with you or to assist others in competing. This clause is important if you don't want your hard work giving free rides to potential competitors.

Usually it is difficult for first-time inventors to secure these provisions, but as your orders to the company increase, so will your leverage. Your initial contracts and nondisclosure agreements can always be revisited.

Designing and Obtaining Prototypes

As noted previously, a prototype is a sample of an invention that demonstrates how the new or proposed product looks and/or works. Prototypes are critical tools for designing, communicating, and marketing your invention. Plenty of ideas

look good on paper, but it is not until the prototyping process begins that the flaws in the inventor's thinking are exposed and so can be addressed. As an inventor, you need to carefully consider how much of your budget should be allocated to designing and producing your initial prototypes and the best way to have them produced.

The Importance of Prototyping

If a picture is worth a thousand words, a prototype may be worth a million dollars. Not only can a prototype help you gauge how potential customers, buyers, or investors will respond to your product, it can also help your attorney better understand your product when drafting a patent application. Most important, the prototyping process can help you identify flaws, inefficiencies, and design alternatives for your invention.

Sometimes a company will place an order for a product as a direct result of seeing a working prototype (when it's supplemented by a clear description of a well-defined market, of course). Allowing the companies to view samples of your product generates excitement. It also shows that you are far along in the development, which inspires confidence in your ability to deliver. The ability to deliver the product, as you will learn in Chapter 5, is a significant factor in the decision-making process of a major retailer to place an order with you.

Types of Prototypes: "Looks-Like" and "Works-Like"

Depending on how you intend to market your invention, you may pursue several courses in creating a prototype. Many inventors are able to make initial prototypes on their own that convey the general idea of how their invention functions, while other inventors opt for a professional-looking prototype that functions as the final product will. The two basic kinds of prototypes we refer to are "looks-like" prototypes and "works-like" prototypes.

A "looks-like" prototype shows how your invention will look to a consumer, which may be adequate if you are planning to license your invention. This kind of prototype is sometimes called a presentation prototype because it's used as a visual aid in presenting a product to a potential licensee. A looks-like prototype is basically a three-dimensional representation of your invention that offers greater visual impact than a drawing would give. There are a number of types of clays and plastic molding products you can obtain in craft stores to begin creating this kind of prototype on your own. You may instead decide to use a professional design firm.

A "works-like" prototype may not be aesthetically complete, but it functions and works just the way the final product will. This kind of prototype is used as a "proof of concept." A "looks-like/works-like" is, as it sounds, a combination of the two and is intended to finalize the operational elements of your invention before it goes into production. The engineering aspects are complete, and the goal of this prototype is to enable you to convince prospective licensors, retailers, and investors that you can provide a product that works.

Designing Your Prototype

The first step in making your prototype is to envision your design and get it down on paper. You should create a simple "requirements" document that identifies the key aspects of your invention, including a description of how the product should look, its dimensions, texture, color, and possible materials. Computer-aided design (CAD) drawings are ideal for this task because they show precise specifications, but for simple products a written description and informal sketch may suffice.

Employing a professional product-development company is the usual route inventors take; however, there are less expensive ways if you are willing to do some legwork. If you need help making CAD drawings (most people do), you can probably find someone to do them inexpensively by getting in touch with a local college or university that has an engineering program. You may also be able to get design assistance there. Keep talking to the industrial design, mechanical or electrical engineering, or computer science departments until you find what you need. Ask if they have a program to assist inventors or can match students with a project. If you don't get it there, call another school. Many colleges and universities now have rapid prototyping equipment and capabilities, which are discussed later in this chapter, and many encourage their students and faculty to assist inventors. It gives the students "real world" experience and, in many cases, puts equipment to use during the year when it would otherwise be idle.

By carefully thinking through and effectively communicating your design, you can minimize misunderstandings with the designer or manufacturer and come up with a prototype that matches your expectations. You may end up with a disappointing prototype if you fail to communicate key specifications.

A Quick Overview of Some Common Prototyping and Manufacturing Methods

There are several ways to go about making a prototype; some are more

Lathes are user-operated devices that can be used to manufacture prototypes.

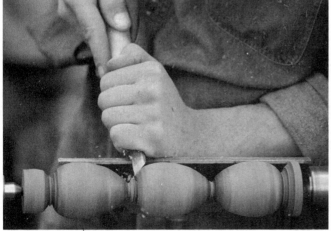

cost-effective and appropriate for a particular invention than others are. The advent of rapid prototyping and the cost of this technology having come down so much in the past few years make this the technology method of choice for many inventors. Here are some old and new forms of prototyping.

CLAY SCULPTING Before the introduction of computer-aided and rapid prototyping, many inventors relied on a block of clay to transform their idea into a visual model. Once the clay was sculpted to "look like" the finished product, the model could be used to assist in the development of tooling to produce the product.

MILLS, LATHES, AND CNC These mechanical devices begin with a block of material (steel, aluminum, wood, plastic, et cetera) and reduce that material using specialized tools until the final object is revealed. (Imagine, for example, a bedpost or a wooden baseball bat rotated and chiseled out of a rectangular length of wood on a lathe.) Mills and lathes are generally operated by a user, whereas computer numerical control (CNC) machines process the work using input from a file sent to the machine via computer.

RAPID PROTOTYPING TECHNOLOGY Rapid prototyping (RP) is a relatively new way to create a prototype. Rapid prototyping machines create physical objects using computer files or CAD drawings that translate the exact physical specifications of an object to create a prototype or model of the object. These machines are not only cost-effective, they can also make your prototypes in far less time that it takes to make a mold. Rapid prototyping can take several days, depending on the method that is used and the size and complexity of your drawings; some rapid prototyping systems can produce models in mere hours. The machine translates a CAD drawing into a thin, virtual, horizontal cross-section. It then creates each cross-section in physical space, one after another, until the model is finished, and fuses those sections together in a way that is not visible to the naked eye. Forms of rapid prototyping include:

Three-dimensional printing (3DP) is a process in which wax is used by a special printer to print a 3-D object. The drawback of this relatively inexpensive process is that the resulting wax object is very brittle and temperature sensitive.

A *stereolithography apparatus (SLA)* uses a laser beam to cure a photosensitive liquid resin line by line, until the final object emerges from the resin bath. This process allows for a highly detailed representation of the product, but it, too, is somewhat brittle and temperature sensitive.

Fused deposition modeling (FDM) is a relatively new evolution of rapid prototyping in which a nozzlelike device squeezes out a melted filament of ABS (acrylonitrile butadiene styrene) plastic

to "draw" the object line by line. The end result is a representative object that is stronger and more durable than the two previously mentioned RP processes.

Before the introduction of computer-aided and rapid prototyping, many inventors relied on a block of clay to transform their idea into a visual model.

Selective laser sintering (SLS) is a process similar to SLA that uses a powdered nylon or elastomer rather than the photosensitive resin. The laser traces each layer, fusing the powder. The advantage of this process is in the variety of materials it can use and the relative strength of the parts.

Rapid prototyping technology became commercially available in the 1980s and was initially used exclusively for making simple models and parts. The machines were expensive, so its use was limited. Today, costs have decreased so much that rapid prototyping can even be less costly than hand tooling prototypes and molds (discussed in the next section). Many businesses and universities now have these machines in their labs.

OTHER METHODS OF PROTOTYPING Here are some common manufacturing processes that, in some cases, can be used to produce prototypes.

Plastic Injection Molding: Chances are, if you are working on an invention made with plastic, you will consider injection molding at some point. The final model of your invention is used to make a steel mold that is likely to be hand tooled by a professional machinist. The model creates a cavity in the steel mold, into which plastic pellets are melted, injected, and then cooled.

Blow Molding: If your invention involves hollow plastic parts, you may consider blow molding. In general, blow molding involves plastic that is melted and extruded, or shaped by forcing it through a die, a hollow tube into which air is then blown to inflate the plastic into the desired shape. For example, blow molding is used to make plastic milk containers, soda bottles, and industrial drums.

Stamping operations for metals: Stamping is used for making metal objects. Metal sheets or strips are pressed using a special tool to form the metal into the desired shape. Stamping is basically a word for applying pressure to metal. Extrusion is a stamping process in which softened metal is pushed through a tool called a die to give it a shape (see below). Other

methods of stamping include piercing, bending, rolling and pressing the metal into tubes and angled structures, embossing, and cutting. Some of these operations require a machine press or stamping press to form the sheet into the desired shape. Often, products are made using a series of these manufacturing processes and machines rather than just one.

Prototypes versus Pilot Runs

A prototype is usually an experimental, substantially handmade, typically one-of-a-kind model of your invention. The cost of a prototype could be hundreds, thousands, or tens of thousands of dollars. Although creating one is often expensive, the value is tremendous in that it allows you to avoid costly mistakes prior to investing in tooling and manufacturing. Preproduction samples are a small quantity of the product that are made as sort of a test run. This pilot run may be used for testing and evaluation of your product or for making samples to provide to customers. In many cases, producing preproduction samples requires an investment in expensive aluminum or steel molds; therefore, the cost per unit is low, but the initial startup is very high.

Be sure to ask prospective manufacturers about their willingness to do small preproduction runs for you economically. Your cost per piece for a small

This T-shaped extrusion press makes window frame parts.

run will likely be higher, and the company's profit margin will be lower, than for a full production run. However, if the manufacturer is looking for a long-term business relationship with you, they will understand the importance of flexibility in making design changes and in having an adequate number of samples to provide to prospective bulk purchasers of your invention, and will work with you on this.

CREATING REPRESENTATIVE PREPRODUCTION SAMPLES

Prior to placing a purchase order for your product, retailers, distributors, bulk purchasers, and other potential buyers want to see a preproduction sample (complete with packaging) that assures them that consumers will be drawn to your idea. At the point when you actually present your sample, it's important that the sample and packaging be as far along as possible, in order to give the impression that you are a knowledgeable businessperson who can actually deliver what you sell.

Of course, prospective purchasers may direct you to make changes in design and packaging before they will enter into a contract to carry your product, so getting the customer's feedback early on in the process is critical in avoiding these late changes to your product.

Making changes to the product or packaging after you have already "tooled up," invested in inventory, and finished packaging design can be extremely expensive. On the other hand, if changes must be made before a retailer will carry your product, this is an expensive lesson that unfortunately needs to be learned. Make sure you factor the costs of accommodating these requests into your manufacturing budget (if you want their business).

SALES-FORCE SAMPLES

Placing that first order to have your product manufactured is a giant leap for your small business. Knowing that the initial inventory, or at least a good portion of it, is already sold makes that decision much easier. Many manufacturers will give you the chance to place an initial order for "sales-force samples" so that you can provide your sales staff or prospective customers with product to evaluate prior to placing a large first order. These samples nearly always cost more to produce than the full run, but in theory, if they help you pre-sell your initial inventory, they can end up helping to significantly lower costs.

MANUFACTURING INITIAL INVENTORY AND/OR INVENTORY TO FILL ORDERS

If you solicit purchase orders, you need to be able to fill them. And a good purchase order from a strong company means investors and lenders will likely help you do that. Payment for this first order could come from the investment that you and/or your investors have contributed. In addition, many inventors are able to finance the manufacturing costs without having funds up front by borrowing against the purchase order using a relationship with a special type of lender called a factor. Another option is having your manufacturer finance the sale by offering you extended terms that allow you to collect from your customer before paying them. Learn more about these types of arrangements by checking out Chapter 7.

> # Whether you are manufacturing overseas or within U.S. borders, you need to make sure your product makes it to the distribution outlet.

QUALITY ASSURANCE/QUALITY INSURANCE INSPECTION Your product is your reputation. Whether you are producing a product domestically or overseas (possibly using a letter of credit), the last thing you want is an end result that doesn't meet your specifications or expectations of quality. Not only is returning the product difficult, but your inability to ship to your customers and the resulting damage to your reputation can be devastating. The first step is to make sure your manufacturer is reputable and capable of producing a quality product. Make sure you thoroughly evaluate their quality standards and get references from their current customers.

Once you are satisfied with the factory, inspecting the product *before* it gets to you is also important. Using an in-country quality assurance company is a good way to guarantee that the quality of the product matches your expec-

tations. In addition, certain types of products may require special testing and certification (see sidebar, page 77).

FREIGHT AND SHIPPING Whether you are manufacturing overseas or within U.S. borders, you need to make sure your product makes it to the distribution outlet. Who pays these costs (buyer or seller) is always a subject of negotiation (see "Understanding Freight and Shipping Terms," page 75.) In addition, as discussed in Chapter 8, if you're manufacturing a product outside of U.S. borders, you'll need to address the costs of shipping the product from the country of origin as well as using a freight forwarding service to get it from its U.S. point of entry to your warehouse. Don't forget to factor these costs into your cost of goods. In many cases, they can be significant.

CUSTOMS AND DUTIES All items entering any country are subject to customs inspection and the assessment of duties and taxes in accordance with that country's national laws. Again, these need to be considered as part of your cost of manufacture.

WAREHOUSING AND ORDER FULFILLMENT SERVICES Inventory takes up space and must be safely stored. Inventory storage involves more than renting a space and stocking boxes floor to ceiling. Inventors who attempt to store and ship inventory on their own soon learn the downside of picking through those boxes to promptly fulfill orders, and keeping track of what's on hand. A good resource for locating a warehousing facility in your area is the Manufacturing Fulfillment Service Association website at www.mfsanet .org. Order fulfillment services are covered in more depth in "Shipping Your Product to Your Customers," page 74. Please note: If you decide to rent space and warehouse the product yourself, you need to carefully insure your inventory against events that can impair it during the warehousing phase.

PRODUCT LIABILITY AND OTHER INSURANCE Most major retailers require proof of product liability insurance. Insurance is generally not a major cost in the grand scheme of things, but you should not consider exposing your business and assets to the risk of being in the marketplace without it. Even if you have established a corporation or other entity to legally limit your liability, you will need this protection (to fend off clever lawyers who may challenge the argument that your personal assets are insulated from personal liability) as your business acquires value. Generally, a few months prior to getting your product on the market, you'll need to begin procuring quotes from product liability insurance providers. This should provide adequate time to have a policy in effect before your first shipment. You'll also want to look into general-premises and umbrella-type business coverage to protect your investment from problems such as fire, theft, and inventory damage. You can even find some policies that will protect you if you inadvertently infringe someone else's patent. If you are using a product fulfillment company, it may cover your inventory. Be sure to resolve this matter by going back and forth between your insurer and your warehouse to see what they cover and what they don't.

CREDIT CARD PROCESSING Credit card processing companies take a percentage of the sales transaction (usually 1.5 to 3.5 percent) to process credit card purchases. Today it is almost a prerequisite to accept credit cards. It is convenient for the shopper, and for you it provides quick access to the cash. Shop around for competitive rates before deciding on a company to process your credit card transactions.

> ### If you decide to rent space and warehouse the product yourself, you need to insure your inventory against events that can impair it during the warehousing phase.

CUSTOMER SERVICE AND SUPPORT/COSTS OF ACCEPTING RETURNS It is unrealistic to expect that any product sold in a retail outlet will not have at least some returned units. In particular, new technology may require that personnel be available to answer questions about the product. Good support can be very cost-effective for you because, by supporting customers and answering their questions well, you may be able to preempt them from returning your product. Most retail purchase orders that you procure will require you to accept returned merchandise, no questions asked, and pay for the shipping. Every time an item is returned, you lose the money. A good customer-service support system (via telephone and/or website) can help you minimize the costs of returns by addressing customers' concerns so they don't return the product.

The foregoing costs (and many more) are also listed in the "Costs to Consider" sidebar (page 60). In order to avoid surprises, become familiar with the checklist and look into which items may apply to your particular product.

Building a Business Case

The term "business plan" has so many meanings that we hesitate to use it here. If it looks as though you can profitably move forward with turning your invention into a product, you need to organize and write down the information that led you to this conclusion and build a business case, a logical argument, for proceeding. There are a great number of considerations to getting your specific invention to market that you have probably thought out, but you need to get the information down on paper. You may find yourself using some form of the document you create to borrow funds, solicit investors, sell your business, or promote expansion. Laying it all out in one organized report allows you to review your assumptions and projections objectively and make sure they coincide with your long-term goals and short-term expectations.

Do not use a business plan form that you get from a book or on the Internet. The people you approach will be financially savvy, and they will focus on how you have addressed relevant issues and whether they can rely on the numbers you have come up with. They will expect anything you provide them to be factual and free of typographical errors. Think of this document as a logical proof of concept rather than as a plan. Facts speak for themselves and should justify the investment. Don't worry about being locked-in long term—all business plans are subject to change.

Relevant information to include in your projections:

- **Revenue:** expected sales per month based on sales cost per unit multiplied by expected units sold

- **COGS:** cost of goods sold; specifically cost per unit to manufacture

- **Gross profit:** the difference of revenue minus COGS

- **SG&A:** the selling, general, and administrative expenses associated with doing business; also referred to as overhead

- **Net profit:** the anticipated profit (or loss) after subtracting your SG&A from gross profit

Once you have these numbers, we recommend that you project them out over a three-year period and also figure projections based on best case, worst case, and most likely scenarios. These numbers will help you weigh the risk versus reward of your idea and justify the financial investment. Your business case should answer the five key questions discussed in this book:

1. What is your product and what makes it unique?

2. Who are your customers?

3. How do you know customers want your product?

4. How much money will it take? (The box on page 60 will give you a quick lesson about costs to consider in developing a profit and loss, P&L, statement.)

5. Where will the money come from?

If you are using your document to solicit investors, explain how much money you'll need to develop your product in various stages and how you plan to use the funding. (Learn more about this in Chapter 7.)

Packaging Your Invention

I t is sometimes said that packaging is the last three feet and three seconds before a consumer makes a decision. Consumers today are faced with such a huge assortment of choices when they purchase a product that getting your product in their hands is no small feat. How will you differentiate your product and get it noticed on a crowded shelf or even a website? Packaging a product is a fascinating mix of artistic, creative, and technological considerations. Shape, size, format, and color are just a few of the details to think about—not just for retail packaging, but shipment or freight packaging, as well. At the outset you'll want to find someone experienced in product packaging and design to help you with your choices. See "Packaging Your Product," page 126.

Getting a UPC

Ever wonder about that black-and-white label on all consumer products that is scanned at the check-out register? In 1972 a group of grocery-industry trade associations formed the Uniform Grocery Product Code Council and worked with a consulting firm to create a uniform product code based on a code system disclosed by an inventor named Ilhan Bilgutay in U.S. Patent No. 3,832,686. Bilgutay created a code based on a simple series of four machine-readable bars of varying widths and put them into groups of two. In 1973 an inventor named George Laurer modified the code for commercial use, and the Uniform Product Code (UPC) was born. The first item ever scanned in a retail establishment was a package of Juicy Fruit gum during a media demonstration. (An entire shopping cart had been ceremoniously loaded with bar-coded items; the gum happened to be the first one picked up by the cashier.)

UPCs were originally created to help grocery stores speed up checkout lines and control inventory; now they're used by virtually all major retailers, which means that you will be required to obtain one if you're launching a product in such outlets. The codes are issued by a private company called the Uniform Code Council (UCC). In order to obtain a code, you must become a member of the council and pay an annual fee to use the code. You can find an online application at www.upc.org. In addition to filling out the online form, you'll be required to provide the following information about your company, which will determine its fee for membership:

A package of Juicy Fruit gum was the first ever item to have its UPC scanned.

- Your company's current revenues

- The number of products you'll be identifying with the UPC symbol

- The number of locations you'll serve

A UPC is sort of a social security number for a product. It appears on products as a series of black stripes with numbers above (see the illustration below for a sample bar code). The code has two parts: 1) a *machine-readable bar code*, which is a series of lines that can be printed at various densities to accommodate a variety of printing and scanning processes; 2) a *UPC number*, which is an accompanying human-readable twelve-digit number.

The first six digits of the UPC number (the prefix) are used to identify the vendor (the supplier of the product—you, in this case). A vendor can use this same UPC prefix for up to 100 products, and you are responsible for making sure the same code is not used on more than one of your products. You are also responsible for retiring codes as items are removed from the product line. The last number of the UPC is called a check digit. It is created automatically by the software to generate the UPC label, and confirms the accuracy of the scan of the previous digits. Creating UPCs is actually quite easy. Specialty software is available to generate the codes quickly after entering the manufacturer code number and product number. Once approved by the UCC, you will be provided with resources to create and print these codes.

Shipping Your Product to Your Customers

As young Earl Tupper learned in 1917, delivering his family's farm produce to people's homes, instead of waiting for them to come to him, was his key to success. Today we are a society that expects immediate gratification. Waiting for weeks, or days, is not acceptable. We want to order and receive our products with the least waiting or expenditure. Large online retailers, like Amazon.com, have invested billions of dollars to shorten the time between order and delivery, and as a result, customer expectations have changed accordingly. In fact, your ability to ship orders reliably and guarantee your delivery dates will be a major factor in deciding whether to sell directly to consumers. It will also be important in whether or not major retailers will feel comfortable carrying your product. As a vendor, your reliable shipping service can set you apart. Conversely, late or partially filled orders can be your downfall, no matter how great your product.

Understanding Freight and Shipping Terms

Who pays the shipping costs for your product? Who owns the merchandise when it has been purchased and is en route to your customer? Both of these questions are subject to negotiation. As an individual inventor or a small company, you might not have the leverage to pass these costs on to the buyer, but it is essential that you understand what freight costs you agree to assume. Before you accept any purchase order from a buyer, make sure you understand the shipping terms, including who assumes responsibility for freight costs and shipment methods, and who determines dates of delivery (since expedited delivery can add substantial cost).

The legal terms for allocating shipping costs are often abbreviated in a contract, and the terms have standard meanings which it is assumed you understand. The most common abbreviation you'll see is FOB, which stands for "free on board." The FOB designation is usually fixed by

a location (for example, "FOB Charlotte, NC" designates that the buyer assumes ownership at the shipping origin—Charlotte, NC—and is responsible for all shipping/transportation costs to get to his or her own facility). Generally, the buyer will want you to agree to the terms "FOB Destination, freight prepaid." However, if you are providing the goods at a lower cost than the buyer could otherwise get them, you may point out that this justifies their assuming the shipping costs and accepting the risk of loss for merchandise in transit.

The following are commonly used shipping terms you may see in a purchase order:

FOB Destination, freight prepaid: This means you, the inventor/seller, pay all the freight costs to deliver the product to your customer/buyer.

FOB Destination, freight collect; FOB Shipping Point, freight collect: These terms indicate shipping costs will be paid by the buyer when the goods reach their final destination.

Freight prepaid and added: This means you (the seller) pay the freight initially, but you add it to the buyer's account when the goods are delivered.

You should be cautious about *cash on delivery* (COD) arrangements unless you have confidence in the customer or have a reliable prior relationship. If they do not honor their commitments, the

Large online retailers have invested billions of dollars to make delivery as fast as possible.

goods will come back to you and you will "eat" the shipping costs, as well as any associated order fulfillment costs.

Finding a Service to Ship Your Product

When you think of shipping companies or freight forwarding services, well-known companies such as UPS, Federal Express, and DHL come to mind. These are reliable companies, and you can't go wrong by using them. However, there are other carriers, including the U.S. Postal Service, that can be competitive depending on your product's size, weight, and delivery location. If you are using an order fulfillment service, they may have arrangements in place with a shipping service that provides significant discounts based on volume. When shipping larger shipments, such as full containers or truckloads, there are many freight carriers that either are options, or can assist you in identifying options.

CREATING A PAPER TRAIL When you receive a purchase order, you should, at a minimum, create documents that show the following:

Order confirmation: Regardless of how you receive orders—mail, Internet, EDI (electronic data interchange), phone, fax—always confirm that you have received an order as well as quantity, payment terms, price, shipping date, and method. Though you should commit to a shipping date, do not commit to a delivery date, as that is not always within your control.

Tracking information: Make sure you maintain tracking numbers for all orders shipped.

A Strategy for Getting Prototyping and Manufacturing Quotes

The goal of this section is to help you find a manufacturer that operates as a true partner in producing your product. A good manufacturer will educate you in the manufacturing process. They will be looking at you as a long-term client; it's worth it to them to see you succeed. Occasionally, a visionary manufacturer will make your first prototype for you at a reduced price because they know that although you're a novice, you'll be back with a big contract if your product is a hit.

When having conversations with a prospective manufacturer, remember that there's more than one way to make a product. Always look at your invention and be receptive to what can be changed to improve its quality and make it more appealing to the future user. Likewise, be open to improvements or changes that can make your invention less expensive to produce. It is helpful to get some input from a manufacturer early on. Your goal is to identify alternative, lower-cost, higher-quality ways to make your product and to ask others to do the same. Manufacturers, whether domestic or overseas, will tend to produce

A Checklist of Questions to Ask a Prospective Manufacturer

Remember that manufacturing costs are only a starting point in determining the overall expenses of producing your product. There are still steps in the distribution chain—shipping, advertising, packaging, insurance—you need to take to get your product to consumers. As previously discussed, the costs to directly manufacture your product should be no more than 25 percent of your retail selling price. This means that you must ask prospective manufacturers specific and pointed questions in order to identify any hidden costs and make sure they can deliver the quality you need.

Here are a few questions to ask, beyond cost per unit, to help you compare various manufacturers who may be considering producing your product.

■ **What types of products do you generally make?** There are many types of manufacturers that specialize in the production of everything from heavy machinery to plastic molded items, chemical formulations to electronics, tools to mechanical devices—to name just a few. To find a manufacturer that has expertise in making your type of product, search www.thomasnet.com or www.alibaba.com.

■ **Are you familiar with quality control standards in my industry and the type of quality control testing that will need to be done on my product?** In the world of manufacturing, experience can translate into savings. A manufacturer who is familiar with your industry, and works with the materials you plan on using, will know the effective shortcuts that don't compromise quality. As an added benefit, they'll probably be familiar with standards and regulations within your industry. In short, simply comparing cost quotes may not give you all the information you need to determine whether a particular manufacturer is valuable to you.

■ **Are you familiar with safety standards for my product or industry?** A manufacturer with this knowledge and experience may be worth more to you than one who quotes you the lowest cost. The savings will be in time and research. However, it is ultimately your responsibility to become familiar with, and independently verify, applicable standards. (Your local reference librarian can be of great help, too, particularly if you live near a large library in a major city.)

(continued on next page)

(continued from previous page)

■ **What types of materials and processes do you most commonly use? What materials and processes would you suggest for my product?** Manufacturers have varying levels of expertise in working with various materials. Moreover, it is possible that your invention may be made from a range of materials (plastic, metals, textiles, rubbers, and synthetic materials). If your invention involves textiles, it's important to ask questions and obtain samples in order to test and determine qualities such as laundering and resiliency. Similarly, if you're working with plastics, it's important to find out about all the types or grades of plastics and the qualities of each (resiliency, cost, biodegradability, dishwasher safety, and so on).

■ **What is your minimum production run?** You may not have the funds for large production runs in the early phases and may not have the capacity to warehouse them. Is the manufacturer willing and able to adapt to your changing needs as you begin ramping up your marketing efforts? In seeking sources for manufacturing, try to estimate the quantities you're expecting to sell as you enter the market, and what your production needs might be if your invention proves to be successful. Ask the manufacturer about quotes for different-size runs (2,000 units versus 10,000 units versus 100,000 units, for example).

■ **How will you charge for samples? How many are you able to provide to me?** You'll find that having good-quality samples is very important to getting initial purchase orders. It is important to know when the manufacturer can provide them, how many, and at what cost.

■ **Do you produce product packaging? What type of packaging do you provide? What is the cost?** You need to be sure to figure the cost of packaging into your total manufacturing costs, and to obtain packaging samples, as well.

■ **What are your payment terms?** Most new relationships begin with payment up front. However, good payment terms may open up a number of possibilities and marketing strategies. Many inventors have been able to launch products without up-front cash by finding manufacturers who are willing to extend credit terms to them. For example, if you get a signed purchase order with a major retailer agreeing to pay you in 30 days, and have 90 days to pay the manufacturer, you can pay the manufacturer directly from the sales proceeds without any other type of loan. (For more, see Chapter 7.)

> ■ **What is your policy with regard to defective products and returned products?** Find out if and how your manufacturer will reimburse you for defective products. This discussion should also include shipping costs associated with any returns and replacements. In addition, try to negotiate with the manufacturer to guarantee that if an agreed-upon percentage (say, 5 percent) of your total order is damaged, they will complete a new production run.

what you give them. If you give them a talking dog dish to manufacture, they will give you back a talking dog dish. They won't necessarily tell you ways in which your dish could be made for half the cost or with improved sound quality, or made more visually appealing for consumers, unless you ask for their assistance. By working with manufacturers who specialize in your industry or product type, you gain from their experience and expertise.

Be sure to ask your manufacturer to help you consider the long- and short-term economics of using a particular process. For example, an aluminum mold may be less costly than a durable steel mold, but if you're going to make a large quantity over a long period, a steel mold with multiple cavities will produce less expensive parts when you calculate the cost of the mold per part. If your goal is just to present your concept, and you're not yet at the mass production stage, it may make sense to forego the mold altogether and go with a rapid prototype (discussed on page 66).

Approach all potential new manufacturers as if they are interviewing to become your partner, because in essence,

they are. You are looking for a partner who will be responsible for producing a high-quality product at a competitive price, and delivering on time. Be selective and make sure it is a partnership that can last.

Finally, be sure that you can explain your product, or that its usefulness is readily apparent from its design and packaging. Consider how your manufacturer can help find creative ways to introduce consumers to your product, using every form of technology available.

Deciding Whether to Manufacture Your Product Overseas

Foreign manufacturing is an important matter for individual inventors and small companies who want to benefit from the cost savings of using overseas labor. International manufacturing is not limited to Asia. Depending on the product, Mexico, Eastern Europe, and other regions offer competitive manufacturing as well.

FROM THE LAWYER
DON'T DISCLOSE TOO MUCH TOO SOON

 Try to form a good relationship with manufacturers early on, so they will want to work with you both in the present, and possibly on future products. But be careful not to give up important legal rights by disclosing too much too soon. Check with an attorney before providing copies of patents, patent numbers, and so on, at this stage. Unfortunately, some manufacturers lull inventors into a sense that a license offer is forthcoming, when really all they want to do is get information to design around the invention or get the inventor's market concept down. It is by no means the norm, but it can happen.

If you are taking a product to market that will compete with products that have been on the market for some time or want to introduce a product at a viable price point, you might have no choice but to begin manufacturing overseas. Regardless of what you ultimately decide to do, you need to be familiar with the full range of global manufacturing options, so you can at least keep up with the competition.

If your product involves a substantial amount of assembly, it may be cost-effective for you to go overseas as you start to manufacture increasingly larger runs of your product. Ideally, all products would be produced domestically, providing jobs and, from a development standpoint, making it easier to manage the factory. In some situations, manufacturing in the United States can be a marketing advantage, and in some cases the "Made in the USA" stamp is required, especially for military products. The realities are that many products end up being produced overseas. We know that this is a controversial issue, but by successfully manufacturing, marketing, and distributing your product in the United States, you create jobs domestically as well as abroad.

Pursuing global manufacturing can seem to be especially intimidating given recent headlines about product recalls that threaten to tarnish the reputations of large U.S. companies. If you choose to manufacture outside of the U.S. (sometimes global manufacturing is the only financially feasible option), you need to know that your finished product will meet all applicable U.S. quality-control standards and will not have hidden defects that cause harm to consumers in any way.

Here are a few guidelines to follow when manufacturing outside U.S. borders:

Find someone local whom you can trust. Unless you are going to be visiting and comparing factories abroad yourself, you'll need to hire a foreign agent that has actual contacts and sources in the country in which you're considering manufacturing. In fact, you should interview a number of foreign agents. You can find them on the Internet, through local trade associations, and by asking companies who they use for their overseas sourcing.

CASE STUDY
Lifetime Achievement

Innovative ideas are often the culmination of life experience. Seventy-eight-year-old Saul Palder was working full time selling catalytic converters when he invented a blockbuster kitchen gadget that has since sold more than 7,000,000 units. When the inventing bug hit, he decided it was time to cut back on his day job.

"One day I was cooking and went to the cupboard to get a container to put some sauce away," Palder said. "The containers fell out, and after I cleaned them up I couldn't find a cover to fit the one I had. I thought to myself, this is crazy." Palder's invention, the Smart Spin, is a free-standing rack that holds twenty-four containers and matching lids, and spins so you can access all sides. The Smart Spin forty-nine-piece storage system, as described on asseenontv.com, "conveniently holds all your storage needs, right at your fingertips . . . yet takes up about the space of a coffeemaker."

Like most inventors, Palder didn't get off to a smooth start, but he stuck with the process: He started with a simple working prototype fashioned from a poker-chip rack, took his hand-drawn sketches to a woodworking shop to make a working prototype, and then actively and persistently pitched the idea himself, taking it to restaurants and food industry representatives. One restaurant owner who rejected the idea noted that *his* problem was getting to the containers at the back of shelves. Rather than get discouraged, Palder went home, added a turntable to his design,

and had a second working prototype made. He also decided it was time to legally protect his market research and design innovations and sought the help of a local patent attorney. "The day we got a patent was probably the proudest day of my life," Saul told the *Boston Globe*. Palder hasn't slowed down since, and neither have sales of the Smart Spin.

Patent drawings for Saul Palder's highly successful modular storage unit.

Stress that you are concerned with quality control. When you begin to look for overseas suppliers, low costs won't be difficult to find. Before you send your product to a factory, though, ask them for samples of their work. In addition, make sure your U.S. representative has actually worked with the factories that are submitting an estimate for your product. Make clear that quality control and accountability are strong priorities that will not be sacrificed for cost.

Send accurate specifications overseas. Sometimes a prototype will be sufficient, but for electronic and similar items it's important to send accurate engineering specifications. You need to have prescribed methods for testing the initial prototype and procedures for verifying that subsequent manufacturing runs will be produced using the same standards as the initial runs.

Have appropriate financial safeguards in place. As your foreign sourcing agent will explain to you, manufacturing overseas is often financed using letters of credit. This is a promise or guarantee by your bank to pay the foreign supplier's bank upon satisfactory fulfillment of the manufacturing contract. In most cases, you have paid the factory by the time the shipment is delivered to your port. It could be thirty days or longer before you see the final product at your doorstep. What if it is the wrong color or size or does not pass your quality inspection? Good luck trying to get a refund. You need to inspect a sample of the final product before it is loaded and shipped and before you pay for it.

Nail down delivery dates and allow adequate time frames. Your reputation is on the line, as well as the future of your business. Getting the first orders was hard enough, but you cannot get paid until you deliver to your customers. Jeff Plitt, a sourcing agent, cautions, "There is nothing more embarrassing to a supplier than poor customer service. Missed delivery dates are often at the center of service issues between supplier and client." Managing expectations is critical. Set realistic deadlines for your supplier and keep in continuous contact as these deadlines approach. By working with experienced and capable sourcing agents, factories, and freight companies, the process can be managed effectively and you can deliver a product to your customer that will, hopefully, lead to more and more orders in the future.

CHAPTER 4

Protect Your Invention

(with as Few Lawyers as Possible)

Have you ever seen those late-night television commercials that promise inventors quick profits for good ideas and urge them to call a toll-free number? Every year "invention promotion" companies bilk inventors out of millions of dollars and often cause them to lose valuable patent rights. Oftentimes inventors (even very smart ones) aren't sure how to get started, and calling the toll-free number seems like a good place to start. It isn't. Many of these companies take loads of money from trusting inventors by telling them they have a great idea and promising to license it in no time. They regularly persuade inventors to enter into one-sided contracts, and then threaten to enforce the contracts against inventors who complain when the companies do not deliver on their promises. While some of these firms occasionally deliver actual results, the odds are one in tens of thousands that the firm will be successful in generating more for your idea than what you pay them in fees. Bottom line: Buyer beware!

FROM THE LAWYER
STEER CLEAR OF SCAMS

 Be very wary of any firm that promises to both obtain a patent for you *and* market your invention. Legal protection and product marketing are very different services, performed by professionals with diverse skill sets and resources. A registered patent attorney or agent knows the law, and has little training in retail pitches or ad copy. A marketing firm is staffed by creatives who know customer markets, not patent statutes.

The U.S. Patent Office has a listing of invention promotion companies that are the subject of complaints, at www.uspto.gov. However, a government website is a rather limited line of defense against companies that have vast advertising budgets and slickly produced commercials. To protect yourself, make sure you do your homework. Research the company online and see if there are any complaints issued against it. You should also ask the company for a list of clients, successful products, and the percentage of clients who actually make money from their idea versus what they pay in fees.

Maintaining Confidentiality amid the Crowds

An important requirement for patentability is that, in general, the invention has not been (and cannot be) disclosed to anyone who is not the inventor more than one year prior to filing. If an invention has been in the "public domain" (in which property rights belong to the community at large) for more than one year, the novelty of the invention is destroyed and the product is no longer eligible for patent protection. However, an exception to that rule occurs when the person to whom an invention is disclosed has signed a special nondisclosure agreement. The U.S. Constitution grants (permits as a right) inventors a twenty-year monopoly on this concept in exchange for full disclosure of the concept. Inventors create something that is novel, nonobvious, and useful, and share with the U.S. (and society) how the invention works, and then, for twenty years from the date of filing, have the right to exclude others from making, selling, or using the invention. In order to ultimately receive this patent protection, the inventor must file for a patent within one year of disclosing or using the idea in public. Because of this one-year requirement, it is critical that you, as the inventor, not destroy the novelty of the idea by offering the product for sale, sharing the idea with

CASE STUDY

Necessity is the Grandmother of Invention

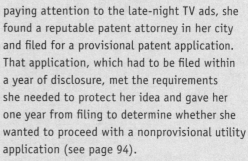

Maria Pistiolis, a grandmother of ten, is a professional seamstress from Charlotte, North Carolina. The concept for her invention came to her one day in the airport when she saw her daughter struggling to juggle a baby, a purse, a diaper bag, *and* a suitcase through a crowded terminal. During the development of her prototype, Maria did not share her idea with anyone as she refined each version, making improvements and adding features. After a week of work at her sewing machine, she presented a rough prototype for a combination purse—complete with bassinet, diaper bag, and changing table—to her daughter to try. As soon as her daughter took her product outside her home, to the mall, to the airport, the clock started ticking. Maria now had one year from the date of "public use" to file for patent protection or she would destroy the novelty of the idea and it would enter the public domain. Fortunately for Maria, instead of

Maria Pistiolis is a grandmother with invention. Her daughter and grandchild were her inspiration.

paying attention to the late-night TV ads, she found a reputable patent attorney in her city and filed for a provisional patent application. That application, which had to be filed within a year of disclosure, met the requirements she needed to protect her idea and gave her one year from filing to determine whether she wanted to proceed with a nonprovisional utility application (see page 94).

Pistiolis brought her patent pending prototype to the *Everyday Edisons* crew, who helped her file patent application No. 2005/0210594. Her patent was eventually issued, and *Everyday Edisons* helped carry her concept to completion as the useful, fashionable bag known today as the Korbie All-in-One Baby Bag. The Korbie features an elegant saddlebag design that offers plenty of space for diapers, bottles, other baby essentials, and babies themselves. And as the baby grows, the removable bassinet can be permanently set aside, allowing the bag to "grow up" with the child and adapt to Mom and Dad's changing needs as parents.

others, or using a prototype anywhere in public or in view of others.

Using a confidentiality agreement or nondisclosure agreement (they're really the same thing) is one way to protect yourself when you do share the idea with other people. While this does not address public use or offering for sale, it does allow you to collaborate with others and get the valuable feedback you need in order to proceed with your invention. Premature disclosure of an invention can jeopardize an inventor's ability to get a patent—which is why the prepatent process has to be handled with such care and is of such importance.

Another reason to steer clear of the purported one-stop invention promotion consulting firms (besides the outrageous price tag of $5,000 to $10,000 for "market research and legal services") is that many of them issue worthless patents that they draft themselves. These patents put your confidentiality at risk because they offer very little, if any, legal protection. Because of disclosures you make to such companies, using these firms may also endanger your ability to apply for a legitimate patent in the future.

Before you even get to thinking about licensing or selling, you have to know that your invention is unique compared with any product that has ever been patented, made, sold, or known to others. Understanding and following the legal requirements of the U.S. patent system is paramount in this process; without that knowledge, you run the risk of falling into sand traps of the legal system and inadvertently contributing your idea to the public domain. Visit www.uspto.gov to learn the guidelines and to search for published patents and applications.

Understanding What Patents Protect

Every good idea deserves a chance. However, an idea must also be able to survive as a product and compete in the marketplace well enough to earn back and exceed the investment in its development. A patent is a tool to gain competitive protection in the marketplace while an invention is being developed, and to allow a true innovator to profit from his or her invention. Patent protection helps drive the development of new products because with it, individual inventors have a way to define and protect innovation.

It rarely makes business sense for an independent inventor to invest in design and experimentation for an idea that the competition can simply copy. Competitors who copy may actually be able to sell the product for less in the marketplace than the inventor can, since they don't have to recoup the cost of experimentation (and sometimes of initial failure) that a true inventor incurs.

Based on the intention of rewarding

Benjamin Franklin, no stranger to the inventing process, helped pioneer the protected rights of individual inventors.

How Much Should I Disclose About My Invention?

This is a recurring question among inventors who enter public forums. They attend clubs hoping to get marketing leads and manufacturing assistance, and may unintentionally forfeit legal options or dedicate their right to make and use their invention to the public domain.

Remember that a public disclosure means telling anyone, other than a coinventor or someone otherwise obligated to keep your invention secret, how to make and use your invention.

Legally, a public disclosure can have implications for your right to file a patent both in the U.S. and abroad. In the U.S., a clock starts ticking after a public disclosure. You must file a patent application within one year of making the disclosure or your invention will be dedicated to the public domain; you no longer have the right to obtain patent protection. Countries outside the U.S. do not have a one-year grace period; if you don't have a pending U.S. application, you lose your right to file a patent outside the U.S. immediately upon making a public disclosure.

In a nutshell, avoid making public disclosures before filing for patent protection whenever possible. There are two ways to do this.

1. Use a Nondisclosure Agreement (NDA)

This is a signed agreement that imposes an obligation of secrecy upon the person you tell about your invention. An NDA is a far more ambiguous form of protection than a filed patent application and can be expensive to enforce should someone breach it. Another drawback is that investors and companies that you approach may decline to sign them. Other times the companies will simply have you sign their own form. Finally, NDAs do not protect you from *all* forms of public disclosure. Specifically, if you

make an "offer for sale," the one-year clock to file starts ticking regardless of whether an NDA is in place. The term "offer for sale" is construed very broadly; telling people your product will be on the market in the future, without pricing or other specifics, may constitute an offer for sale.

2. Safely Make "Non-enabling" Disclosures

Sometimes you may want or need to communicate about your unprotected invention, but it is not practical to get an NDA (e.g., having everyone at your local inventor's club meeting sign one). If this is the case, you need to make sure your disclosure is not "enabling." An enabling disclosure is one that gives someone enough information to make and use your invention.

Generally, it's okay to tell people your field of invention ("consumer electronics" or "software"). It's also safe to discuss the problem you are working on solving ("saving energy" or "a new consumer product"). You can also provide your target market ("business users" or "teens"). This gives people enough to get a sense of your goals, but is not generally enough information to constitute a public disclosure and affect your patent rights. It isn't complicated, but it does require that you plan before you speak about your invention. Here are examples:

A non-enabling disclosure: "I'm working on an electronic device that promotes weight loss."

An enabling disclosure: "I'm working on a refrigerator handle that produces an electric shock."

When in doubt, ask an attorney to help you describe your invention in non-enabling terms. One of the great advantages of filing a patent application is that the patent pending status gives you more freedom to talk about your invention and to market it.

FROM THE LAWYER
PUBLIC ACCESS

Under most circumstances, patent applications are required to be published by the USPTO within eighteen months of the date on which they are filed with the patent office. Once your patent application is published, your competitors and any other member of the public can access it freely on the USPTO website at www.uspto.gov and learn from the information you have disclosed. They are free to attempt to *design around* your patent and improve upon your invention, thus paving the way for further advances and improvements that benefit society.

the individual inventor, patents offer legal protection to ensure that inventors are fairly protected while developing their ideas. Our Founding Fathers were so well aware of the societal importance of independent innovators (such as Benjamin Franklin, who invented bifocals, lightning rods, and other hot Colonial consumer items) that they wrote it into the Constitution: The original document (Article 1, Section 8) specifically authorizes Congress "to promote the Progress of Science and useful Arts, by securing for limited Times to . . . Inventors the exclusive Right to their . . . Discoveries" and lays the groundwork for a protective environment for future innovation. Patents are intended to encourage technological progress by giving the true inventor a head start in the marketplace—they stave off competitors for a set amount of time so that inventors can recoup the costs of creating, developing, and distributing their product.

How Patents Empower Individual Inventors

Ultimately, patents give inventors power. Whether you're a multinational conglomerate with two dozen facilities or an individual inventor working out of your garage, a patent gives you a "mini monopoly" for the entire term of the patent. You can prevent others from making, selling, or using your invention (or its functional equivalent) for a twenty-year period from the date the patent application is first filed. A United States patent provides protection throughout the U.S. and also prevents the importation of infringing products from anywhere in the world.

You can also file for foreign patent protection to obtain protection in other countries. Every country in the world exercises control over patent protections recognized within its borders (see sidebar "Getting Global Patent Protection," page 91). Before exploring additional patent protection, you need to determine whether, based on your earlier analysis, there is a market for your product in foreign countries. Foreign patents are expensive, so make sure the economic returns justify the investment.

Patents protect all functional equivalents of your product that produce the same result as your invention in the same way your product does it. This is a powerful principle of patent law known as the *doctrine of equivalents*. This doctrine is just one of many reasons patent protection is so valuable.

Who Is a "Person Having Ordinary Skill in the Art"?

United States patent law is full of references to a hypothetical "person having ordinary skill in the art." We will refer to this person as Ms. P-H-O-S-I-T-A.

Ms. Phosita is a very important person in U.S. and foreign patent law, despite the fact that she doesn't actually exist. Legally, the disclosures contained within your patent application must be sufficient to enable this "ordinary person" to make and use your invention. Ms. Phosita is a legal fiction considered to have the normal skills and knowledge of a person in a particular technical field (for example, plastics manufacturing or biomedical engineering) without being a genius. Ms. Phosita is a reference for determining whether your invention is nonobvious to a person of her skill level or involves an "inventive" step as required in Europe.

In both the U.S. and Europe, Ms. Phosita's skill level, background, and education differ according to the technical field in which the invention is being evaluated.

The better your patent application is written, the more marketable variations of an invention it covers. Different versions of the same invention are called *embodiments*. For example, a patent on a feature of a digital music player that allows you to instantly change song tracks may cover several incarnations of the same technology. One embodiment of the invention may be a car stereo system, another embodiment may attach to a stroller, and still another may be adapted for joggers. Assuming the patented aspects remain the same, the patent will bar competing products that make aesthetic or other obvious modifications that don't change the patented function of the product (in this case, rapid song-track shifting). See Appendix C for a sample patent; various sections of the patent are identified to help you understand the mechanics of patent drafting.

While there are a lot of commercial companies that offer "patent services," patents are ultimately legal documents and subject to legal interpretation. Legal interpretation of the language in your patent can be very important over its life, particularly if issues of infringement ever crop up. Patent attorneys have to pass a special examination, be

FROM THE LAWYER
NONPUBLICATION REQUEST

Inventors can file a special nonpublication request for a *pending patent application* with the USPTO if they file a special form and agree not to file a foreign patent on the invention. However, your patent will still be published once it is issued. There's no way around it.

registered with the USPTO, and have a technical or engineering background. The patent office also allows individuals called patent agents, who are not attorneys, to take the patent exam. Patent agents can file documents with the agency, but cannot give legal advice. The USPTO maintains a list of registered patent attorneys and agents at www.uspto.gov/OEDCI. You can search the website by city or state to find registered patent attorneys or agents in your area.

FROM THE LAWYER
NONDISCLOSURE AGREEMENTS

 Avoid telling anyone about your invention without having them sign a nondisclosure agreement (NDA). (See Appendix A for a sample NDA.) Most foreign countries don't have a one-year grace period for filing a patent; thus *any* disclosure of your invention to someone who has not signed an NDA can permanently bar you from filing a patent in these countries.

The Real Price You Pay for a Patent

Aside from the legal fees you pay your attorney, patent protection has another price. You must actually disclose important information about your product to your competitors. In exchange for receiving a patent, the law states that an inventor must disclose how to make and use his invention in sufficient detail to allow someone of "ordinary skill in the art" to replicate it. This information is essential. This is a trade-off: In exchange for sharing your invention, you get twenty years to prevent others from making, selling, or using it. However, when the patent term expires, the invention immediately falls into the public domain and can be used by anyone.

If your invention, and the process whereby it is created, would be difficult if not impossible for others to figure out, it may make sense not to file a patent and to rely on a trade secret instead. The formula for Coca-Cola was never patented. Had a patent been filed, the recipe would now be available for anyone to use. Instead, the recipe has been a closely guarded secret, and as a result, no one has figured out exactly how it is made. As an inventor seeking a patent, you need to fully disclose the *best mode* of making your invention, and you must make sure that the instructions you include in your patent are sufficient to explain it to someone of ordinary skill in the art. Refer to the box on page 89 to find out more about the legal standards that define the characteristics of this hypothetical person.

Patents are, in essence, a contract between the government and the patent holder. To encourage private entrepreneurs to reveal their research, testing, and experimentation *and* to make it available to others, the government gives the inventor a twenty-year monopoly. When the monopoly expires, the invention enters the public domain. That means the intellectual property rights now belong to the public at large,

Getting Global Patent Protection

Once you've filed your United States patent application, your next step will be to decide whether your invention has the commercial potential to justify filing for protection in foreign countries. You must file for foreign protection within one year of the date you filed any type of United States patent application (including provisional applications). If you are unsure of where you'll ultimately want to seek patent protection, you can still buy some time. You can file a patent application under the Patent Cooperation Treaty (PCT), reserving the right to seek protection in any country that is a party to the treaty, without having to narrow down your choice of country until thirty-one months from the date of filing the application in the United States.

Dozens of countries (including most of Europe) are parties to this treaty. Keep in mind that most foreign countries require that you file for patent protection before making any enabling public disclosure (which anyone could use to produce your invention on their own before your patent is published). This process is very expensive, so make sure the business case you build shows that such expenses are justified.

and anyone can freely make, use, sell, or profit from them. The United States patent system is a vast repository of more than two hundred years of technological information (called *prior art*) that is an indispensable reference to contemporary innovators.

Congress presumes that if inventors publicly disclose their invention, one year is plenty of time to seek a patent. United States law specifies that one year after the date of the public disclosure, the incentive to give a patent in exchange for the information no longer exists, since people presumably already know about it. Therefore, it is critical that you avoid disclosure or public use, and refrain from offering your products for sale, until you are ready to seek patent protection.

Some Legal Landmines All Inventors Face

Most inventors find the patent process intimidating the first time through, but by the time they apply for their second or third patent, they know what to expect from the system and what the system expects from them. Here's a brief overview of the major legal landmines you can expect to encounter.

■ *You must* invent something that meets the legal standards of being "novel" and "nonobvious." Your invention must be novel: different from anything else that is patented or is otherwise

(continued on page 94)

Documents to Protect Your Invention

INVENTOR'S NOTEBOOK
It's important to keep a record of when and how you came up with an idea, and the dates you disclosed it to others. Your inventor's notebook (which can be as simple as a file on your computer, or an elaborate diary of each developmental stage) is a detailed record of what you disclosed to whom on what date, depending on the level of secrecy you feel is necessary.

Inventors' notebooks have persuasive value if there is a dispute about who is the actual inventor of an invention. Although many countries have a "first to file" system, only the true inventor has a right to file.

NONDISCLOSURE AGREEMENT (NDA)
If you disclose an invention to anyone other than a coinventor—including your spouse—you can lose your international patent rights and significantly limit your time frame for filing for a patent in the United States. The date on which you "publicly" disclose your invention to someone other than an inventor is known as the "critical date," after which you have one year to file for patent protection in the U.S. Foreign countries do not have this grace period. In the U.S., if you disclose your invention under an obligation of secrecy, such as that imposed by a signed nondisclosure agreement (like the one in Appendix A of this book) or in communication

with an attorney, the one-year clock will not start running and will generally preserve your foreign filing rights.

The NDA can help you preserve your filing rights that you would otherwise lose by talking publicly about your invention. As a practical matter it may be difficult to enforce, so do not view it as a fail-safe form of legal protection against companies that have a poor track record honoring them.

NONCOMPETE CLAUSE
This is a clause that sometimes appears in an NDA or may be a separate agreement. It defines that those to whom you disclose your invention won't use your proprietary information to compete with you. If you can get companies in your industry to sign this, it's to your advantage. Fair warning, however, that most of them won't.

LETTER OR E-MAIL DOCUMENTING THAT YOUR PRODUCT PRESENTATION IS NOT AN "OFFER FOR SALE"
The "offer for sale" issue is one that catches many inventors by surprise. Not only does a public disclosure mark a one-year timer, so does any offer to sell your product. It does not always matter if you have an actual product; sometimes a brochure about a future product can trigger the one-year timer. Thus, it's a good idea to send a letter or e-mail in advance of a meeting confirming that you do not intend your presentation to be an offer for sale.

Better yet, include this provision in your nondisclosure agreement.

PROVISIONAL PATENT APPLICATION

This is a temporary type of application that affords you the "patent pending" status. When you are serious about presenting a product but don't want to get bogged down trying to convince companies to sign an NDA, a provisional patent application offers better protection than an NDA, and it affords you that patent pending status. Generally, a lawyer will prepare a Provisional Patent Application for around $2,000. You'll need to turn your provisional patent application into a nonprovisional application within one year of filing it, or it will become abandoned. If your product looks promising, convert your application to a nonprovisional sooner so that the patent office will begin examining it. Do not file a provisional application too early in the course of doing your full market research, unless you are ready to file a nonprovisional application with formal claims and make decisions about foreign filing within one year of filing it. However, once you are reasonably confident of your market and have a sense of how to produce your product, this is an important step to take.

If you have done your market research and know that it will be in your interest to move toward protecting a profitable product, a provisional patent application will provide you greater protection and more flexibility in marketing your product than an NDA and gives you the right to put "patent pending" on your product to deter others from entering your market space.

NONPROVISIONAL (UTILITY) PATENT APPLICATION

This is a formal patent application with claims that is ready to be examined by the U.S. patent office. Once your application is examined and issued, you have a legal right to demand that infringing competitors take their products off the market. As of the writing of this book, a Nonprovisional Application for most consumer products will cost you about $3,000 to $5,000 in attorney fees (more or less depending on the complexity of your invention).

This document gives you the right to exclude others from making, selling, or using your patented invention within the country for which the patent protection applies.

AMENDMENTS AND ADDITIONAL APPLICATIONS TO PROTECT MODIFICATIONS

If you significantly change your invention after filing your initial application, be sure to check with your attorney to see if your application claims, as originally drafted, still cover what you are marketing.

Applications and modifications to pending applications can help obtain the broadest scope of protection possible as you continue to develop your product.

(continued from page 91)

known to others anywhere in the world. In addition, it's not enough to just change what's already been invented; you must change it in a way that is truly new and therefore not obvious.

For example, in May 2007 the U.S. Supreme Court ruled in *KSR International Inc v. Telefex* that simply adding an electronic sensor to a different location on a gas pedal to make it compatible with a particular vehicle was an obvious modification of an already patented gas pedal. Obviousness is a subjective standard that lies in the eye of the beholder (the patent examiner). It's a topic that patent attorneys argue about endlessly. Because the concept of obviousness is controversial in just about any context, most patent infringement cases in which the issue is raised end up being settled out of court. Who wants to take a chance on guessing what a jury will consider obvious? (The obviousness standard is covered in more detail later in this chapter.)

■ *You must* be able to prove that you are the true inventor and that all the inventors are named on the patent application. Your invention cannot be known to anyone else anywhere in the world. For example, if you see inhabitants of a remote tropical island using a specific plant for a specific purpose, you cannot go back home and patent it. You are not the true inventor, even though the invention has never been patented or seen in the United States. Also, everyone else who is an inventor of the item must be named on the patent application as a coinventor, otherwise they can sue to have their names added later and enjoy the same patent rights. If you have other people (such as engineers) advising you, you may need to name them on the application as coinventors. However, you can have them sign an agreement that says they assign their economic rights in the invention to you, which is a common practice. Many big companies do this with the people who work for them doing research; they have them assign their invention rights to the company as a condition of employment.

■ **Be careful not to sabotage yourself by telling others (even family members) too much about your invention too soon.** You could inadvertently blow your patent rights if those people haven't signed a nondisclosure agreement. If you do blurt out information about your invention, you need to do some damage assessment. You need to figure out if you've actually made an *enabling disclosure,* which means giving someone enough information about your invention that they could make it on their own without waiting for your patent to be published. Once you make an enabling disclosure to someone who hasn't signed an NDA, a clock starts ticking. You must file a patent application within one year of the enabling disclosure or your invention enters the public domain.

■ **Be careful not to promote your invention too soon.** The rule that you must file a patent application within one year of offering your invention for sale is called the *on-sale bar* and is triggered by anything that can be construed as offering your invention for sale. This includes promoting it at trade shows, accepting purchase orders for it to be filled at a future time, or trying to license the concept to a manufacturer. Courts have interpreted a wide range of activities to be offers for sale, such as accepting advance orders when a product is not yet invented or handing out literature about future products at a trade show. Note that although a nondisclosure agreement protects you in the event that you make an enabling disclosure, it offers you no protection from the on-sale bar. If you offer your invention for sale, even if the other party has signed an NDA, you must file for patent protection within one year of the offer.

■ *You must* be the first inventor to file a patent application on the invention in order to protect the full range of your rights. Historically under U.S. law, patents have been awarded to the first person to invent. The USPTO currently has some cumbersome and expensive legal procedures in place to figure out who is the first inventor when more than one files an application on the same invention. As a practical matter, if you plan to establish patent rights for an invention, you should file an application as soon as possible. Since its inception, the U.S. patent system has awarded a patent to the first person who invents

the product, rather than the first to file for a patent. This "first to invent" versus "first to file" is a controversial issue, since most countries use "first to file" as their procedure. You cannot go back in time and claim you came up with an idea (and didn't file a patent) before someone else; you can, however, argue that if you

FROM THE LAWYER
A "RIGHT TO EXCLUDE"

It was explained earlier that a patent protects you from others' infringing on your invention. A patent *does not* give you the right to sell your invention. Rather, it *prevents others* from selling it. What is the difference? Although you can get a utility patent on an improvement for an existing product, you might not have the legal right to make or sell your invention until you get a license from the inventor of the product you've improved. (The inventor of the underlying product retains the original rights regardless of your improvement.) Without a license, you may be infringing on the original invention rights if you attempt to produce yours. Often an inventor will approach a company that owns the rights to a product to see if that company will license it for improvement. If you opt to do this, it's a good idea to file a simple provisional patent application to protect your rights and your negotiating position with the established company. Many companies will not sign nondisclosure agreements, and regrettably, even when agreements are signed, individual inventors can rarely afford to sue in order to enforce. an NDA in place, you should not consider it a complete defense against companies that have poor reputations in dealing with individual inventors. Fortunately, most companies that sign NDAs do so with the intention of honoring them.

and someone else filed patents, you can document your invention's having been created first. This can be accomplished in many ways, including the use of an inventor's notebook that contains dated entries and of witnesses to the original date of inventorship. Be aware, however, that many experts feel it is simply a matter of time until the U.S. changes to a "first to file" system in order to bring U.S. patent law in line with the rest of the world.

■ *You must* go through the process of prosecuting a patent application with the United States Patent and Trademark Office. Your patent application will undergo a rigorous examination at the USPTO to determine whether it is indeed novel and nonobvious, and whether your patent application sufficiently teaches others how to make and use your invention. This is known as the patent *prosecution* process. It can take a couple of years (or longer) to prosecute your application; but fortunately, the minute your application is properly filed with the Patent Office, you can use the "patent pending" designation to put the world on notice that you intend to establish and enforce your patent rights. If someone makes or uses your invention and your patent application was published at the time, you may be able to collect royalties from them once your patent is issued.

A Sample Patent Claim

Here's an example of appropriate and specific claim language used to describe a light-up flying disc in a patent application.

WE CLAIM:

1. A flying saucer toy which comprises a generally disc-shaped body terminating at its periphery in a downwardly pointing rim, said body and rim defining a generally convex upper surface and a generally concave lower surface; power source retaining means generally centrally located on the underside of said body; lighting means generally fixedly operatively positioned proximate said rim and visible when energized from the outside of said rim; means for holding said lighting means in its said operative position; said lighting means holding means comprising at least one arm extending generally radially from said power source retaining means to said operative position of said lighting means; and electrical circuit means extending from said power source retaining means to said lighting means for conducting electricity from said power source retaining means to said lighting means so as to energize said lighting means.

What Can You Patent?

The U.S. Constitution created the patent system, authorizing protection for "useful arts," in the eighteenth century. The term *patent* was originally interpreted to refer only to processes, machines, and manufactured goods. These types of patents are now known as utility patents. Since then, patent law has evolved and expanded to allow inventors to patent ornamental designs, business methods, and even certain types of plants. The courts have also decided that computer programs, business methods such as priceline .com (in which you name your price for travel), online ordering systems, and even some life-forms (such as stem cells) are subject to patent protection.

Getting Basic Patent Protection

Although U.S. law provides for several types of patents (for example, patents on plants and on designs), a *utility patent* on an apparatus or method is the most common type of protection and offers the broadest form of coverage—it is the gold standard of patents. (It's the type of patent that inventors featured on the *Everyday Edisons* show obtain with the help of the program's legal staff.)

Basically, a utility patent covers anything that functions in a unique manner to produce a unique result.

FROM THE ENTREPRENEUR
MAKE YOURSELF "USEFUL"

Consider, for a moment, the "usefulness" requirement strictly from a marketing perspective. To get a patent, your invention must be deemed useful by the reviewers. In addition, it must meet this requirement at retail in order for it to sell to consumers. A perpetual motion machine can't be granted the "useful" label because it won't work—and that is also the reason it wouldn't sell. Above all, your invention has to work and have some practical application. After this, everything else will fall into place.

Examples of utility patents include: No. 3,570,156, for a lava lamp (see next page); No. 5,960,411, for a "method and system for placing a purchase order via a communications network" (for the now famous Amazon .com "one-click" method of Internet shopping for Amazon's products); and No. 5,851,117, for a method of training janitors.

A utility patent contains special language, called *claims*, that describes the features and elements of an invention that distinguish it from competing products. See the box on page 96 for sample text from a claim in U.S. Patent 3,786,246, for an "illuminated flying saucer" (basically a light-up Frisbee).

Patent attorneys carefully choose the words they use in claims. They choose words that have the broadest possible meaning and then try to make them even broader. For example, in the claim language in "A Sample Patent Claim," instead of saying "a saucer," the attorneys

March 16, 1971   E. C. WALKER   3,570,156
                         DISPLAY DEVICE
                      Filed Nov. 13, 1968

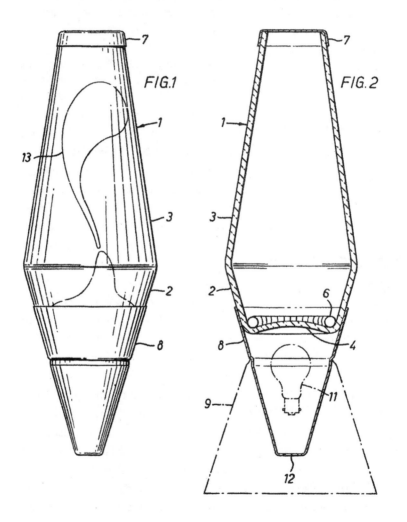

At its peak, Craven Walker's Astro Lamp (commonly known as a lava lamp) sold seven million units a year worldwide.

use the words a "generally disc-shaped body" to encompass anything that is even remotely shaped like a disc (even if it has ridges or bumps or scalloped edges). They also use "electrical circuit means extending from said power source" instead of just saying "battery," in case the competition comes up with a light-up Frisbee powered by something other than a battery during the twenty-year patent term.

The more claims you include in your patent, the more legal protection your patent gives you. (Appendix C, page 213, takes you through the process of reading an actual patent and gives you tips for figuring out what a patent claim really covers.)

Utility patents are the broadest and most common form of patent protection. To qualify for one, your invention must be 1) useful, 2) novel, and 3) nonobvious.

Meeting the "Usefulness" Requirement

To be deemed useful, your invention must have a stated purpose and must actually work. Fortunately, usefulness is usually the easiest criterion to meet, because most inventions are developed to serve a need. The use can be purely aesthetic, such as waterproof mascara or a fabric that doesn't wrinkle. There are a few things that the USPTO will, however, reject on grounds of "usefulness" (or lack thereof) primarily because they don't work. For example, perpetual motion machines are routinely rejected because they defy the laws of physics. Many examiners classify inventions that

FROM THE LAWYER
PROFESSIONAL PUBLICATIONS

Whether you're a scientist, an entrepreneur, or a student, you may be eager (and understandably so) to publish your new findings. Doing so can earn you a lot of well-deserved recognition in your field. Unfortunately, it can also operate as a bar to your patent if publication occurs more than one year prior to the filing of your application.

purport to run indefinitely without a rechargeable power source as perpetual motion machines.

Establishing Novelty

Novelty is the requirement that your invention be unique compared to any prior art. Prior art, as noted previously, includes existing patents and products that other people have already envisioned. It includes publications anywhere in the world, and technology that has been used privately and has been known about even if it was never introduced into the marketplace. Sometimes inventors create their own prior art by writing about their inventions (for instance, presenting them as a thesis or on a blog) before seeking patent protection. It's always a shock for inventors when patent examiners cite the inventors' own publications against them as prior art. The takeaway: Patent, then publish.

Overcoming Obviousness

In addition to proving that your invention is novel, you also have to convince a patent examiner that it is "nonobvious." This means that at the time you came up with your invention, it could not have been evident, without much thought or experimentation, to people who are "ordinarily skilled in the art" to do the same thing.

In particular, patent examiners are on the lookout for patent applications that appear to claim a combination of things that are already known in the art. For example, e-mail programs are well known and so are digital music files. A device combining an e-mail server and digital music files would be obvious. Similarly, a cell phone with a holographic image that corresponds to a ring tone to create a "theme phone" is probably not a patentable concept because holographic images are known in the art of telephony and so are cell phones and ring tones. Most patent examiners would likely reject the theme phone as an obvious combination of existing elements.

Of course, any invention taken in hindsight is arguably obvious. It now seems normal to stick a metal rod in the air to divert lightning; but no one was rushing out in the rain to do it in Benjamin Franklin's day (except Ben, that is).

Similarly, Thomas Edison's patent for a lightbulb (U.S. Patent No. 223,898) consisted of a combination of known elements (thin carbon filaments, glass bulbs, platinum wires) and predictable interactions. There was also an abundance of light-emitting objects, so it might seem obvious to channel a predictable result for an activity commonly engaged in by consumers these days. At the time, however, few people questioned whether Edison deserved a patent.

Today there are many arguments you can make to a patent examiner to overcome an obviousness rejection. One position you can take is that your invention solves a problem that the

Sometimes a Picture Is Not Enough: Limitations of Design Patents

One brilliant engineer learned the difference between design and utility patents all too well and, ultimately, too late. The inventor showed a company representative the type of airplane propeller he had developed, and based on several discussions and a viewing of the prototype, the company was eager to license the invention. It functioned differently from anything out there. The prospective licensee asked to see his patent, and when they found it was a design patent, regretfully had to back out of the deal. The design patent didn't give the inventor any rights valuable enough to license the product—anyone (including the company's own engineers) could make a trivial modification to the design and avoid patent infringement. The inventor, in all likelihood, could have obtained a utility patent on the functionality of his invention.

competition did not solve, and that commercial success has proven the value of your inventive contribution. Your patent may still be pending by the time you take your product to market and any commercial success of your invention may be cited by courts as evidence that an invention is nonobvious.

In May 2006, as mentioned, the U.S. Supreme Court released its decision in *KSR International Inc v. Telefex*, an important case on the matter of obviousness that addresses which ideas are entitled to protection under the U.S. patent system. This landmark case makes it harder for inventors to prove that their inventions are not obvious. Simply adding a part to an existing invention is an obvious change. In the *KSR* case, the Court ruled that "granting patent protection to an advance that would occur in the ordinary course without real innovation retards progress and may, in the case of patents combining previously known elements, deprive prior inventions of their value or utility."

Needless to say, obviousness remains a subjective determination and a hotly debated issue within both the USPTO and the court system.

Design Patents and Their Downsides

Design patents, as compared with utility patents, were introduced in 1842 to protect original and ornamental designs for articles of manufacture. They are generally more limited, and blur the line between copyright law (which has traditionally protected artistic expression) and patent law (which has traditionally been used to protect functional ideas). A well-known example of a design patent is for the Statue of Liberty, patent number D11,023, awarded to August Bartholdi in 1879.

Design patents have, in recent years, been abused by many of the fly-by-night "invention promotion" companies about which the USPTO warns inventors. These companies take advantage of inventors by offering cheaply prepared design patents and charging top dollar for them, when the inventors really should have been filing for utility patents. Design patents are the simpler and cheaper ones to prosecute in the USPTO, because they contain only one claim: for the *exact* ornamental design and appearance of the product shown in a picture included with the patent. If your competitor makes even a minor modification to the appearance of the product, it may be very difficult (or impossible) to enforce your patent in the marketplace. This is not to say that design patents provide no protection or have no value. If your invention has an iconic look to it that others would want to copy, then a design patent is useful. Also, the use of a design patent with a utility patent provides greater protection to inventors, enabling them to protect the function and look of the product. The Apple iPod, for example, has become one of

The iconic Statue of Liberty was protected using a design patent.

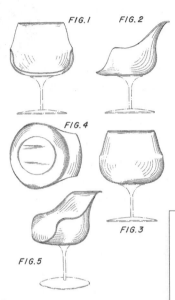

United States Patent Office

Des. 205,939
Patented Oct. 11, 1966

205,939

CHAIR OR THE LIKE

Erwine Laverne and Estelle Laverne, Washington Square
Village, New York, N.Y.

Filed Jan. 4, 1962, Ser. No. 68,190

Term of patent 14 years

(Cl. D15—1)

FIG.1   FIG.2

FIG.4

FIG.3

FIG.5

FROM THE LAWYER
DESIGN VS. UTILITY

Design patents have a life span of fourteen years, as opposed to utility and plant patents, which last for twenty.

These are some of the drawings the court used to compare the Saarinen chair (right) and the Laverne chair (above).

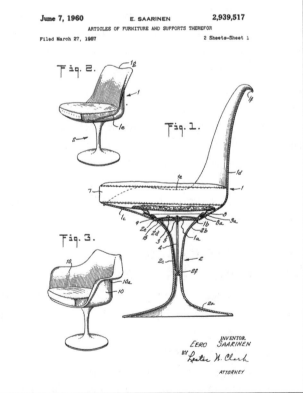

June 7, 1960   E. SAARINEN   2,939,517
ARTICLES OF FURNITURE AND SUPPORTS THEREFOR

Filed March 27, 1957   2 Sheets—Sheet 1

Fig. 2.

Fig. 1.

Fig. 3.

INVENTOR.
EERO SAARINEN
BY Lester N. Clark
ATTORNEY

the hottest-selling personal electronic devices of all time. Apple's intellectual property includes utility patents (such as the click wheel) as well as a design patent on the iconic look of the product. While there are many other MP3 players available, none can legally function and look like the iPod.

The perils posed by relying on design patents to protect an invention are illustrated by a famous case involving an enormously popular type of chair. In 1958 Eero Saarinen developed the famous pedestal chair, which consisted of a single piece of molded plastic that formed a chair seat, which was supported by a tulip-shaped pedestal rather than traditional legs. Mr. Saarinen carefully experimented with various materials and dimensions to make sure the chair would support the weight of the user. For some reason, despite the functionality of the pedestal and his significant investment, Saarinen opted to file a design patent rather than a utility patent.

In 1968 another designer, Erwine Laverne, applied for a design patent for a pedestal type of chair with a slight difference in the shape of the molded seat. The U.S. Court of Customs and Patent Appeals, relying on the drawings produced by the USPTO (shown opposite), sided with Mr. Laverne and granted him his own patent for the chair. Had Mr. Saarinen gotten a utility patent, Mr. Laverne and countless other competitors would have found it far more difficult (if not impossible) to capitalize on the unique pedestal chair

and all the experimentation that went into making it work as intended. That said, Saarinen continues to be known as the originator of this particular style of chair; the colloquial term for any chair resembling this style, regardless of its originator, is "Saarinen."

Patenting Plants

In 1931 the Supreme Court upheld a patent for Henry Bosenberg's climbing, ever-blooming rose. This was the first time the Supreme Court had acknowledged the now well-established principal that an inventor of a plant is the first person who "discovers" and reproduces its distinctive qualities by breeding or grafting.

Plants that already exist in a wild or uncultivated state can't, of course, be patented, because they occur freely in nature, not as the product of some sort of scientific discovery. New and distinct varieties of Kentucky bluegrass are examples of "manmade" patented plants whose breeding does not happen naturally (without the aid of human involvement).

The protection afforded by plant patents is the exclusive right to reproduce the plant. They are easier to obtain than utility patents, but they provide less protection. Holders of plant patents can't protect themselves against someone purchasing patented seeds and then (legally) selling the plants they've grown. Holders of utility patents can, however, claim infringement if a plant covered by the patent is sold without permission.

Requirements for plant patents differ slightly from the requirements for utility patents. The requirement of "distinctiveness" is substituted for the requirement of "utility"; instead of novelty, utility, and nonobviousness, plant patents require novelty, distinctiveness, and nonobviousness.

Patents That Protect Software Programs

Albert Einstein wouldn't have been able to patent his formula, $E=mc^2$. The formula is an idea, a theory about a law of nature, and not a specific device or process. Patent law draws a sharp distinction between abstract ideas and specific applications. Ideas, such as Einstein's theory of relativity or Newton's law of gravity, are considered unpatentable laws of nature.

Until the 1980s it was virtually unheard-of to get patent protection for computer software. Prior to that time, courts uniformly held that computer programs were types of mathematical algorithms and, since mathematical algorithms are laws of nature, patents were not appropriate. Nevertheless, software inventors are clearly among those whom the constitution was intended to protect.

In 1981 the U.S. Supreme Court opened the door for software patents. In a case called *Diamond v. Diehr*, the Court characterized a computer program for regulating the temperature in rubber molds as a process rather than an algorithm. The Court reasoned that the "process admittedly employs a well known mathematical equation, but [the developers] do not seek to preempt the use of that equation. Rather, they seek only to foreclose from others the use of that equation in conjunction with all other steps of their claimed process."

Should You Patent Software and Business Methods?

Many companies around the world view a business method patent as a good defensive and competitive measure. That's because this kind of patent may make it harder for a new competitor to enter their market using their methods, and a patent could be an asset worth showing on their balance sheet. A business method patent could also be the foundation of a completely new type of business. Take, for example, Priceline.com, the online service that allows buyers to name their own price for travel. This company filed for, and received, numerous patents on the business method of allowing buyers to name a price and sellers the ability to accept or decline the offers.

The patents that were issued to Priceline.com gave the company the ability to prevent others from entering this market, and allowed them to establish their brand identity without dilution from competition.

In 2008, the United States Court of Appeals for the Federal Circuit (CAFC) upheld a decision of the USPTO rejection of patent claims involving a method. The court said that a method, such as the steps in a software program, qualifies only if the method is 1) implemented with a particular machine (one specifically devised to carry out the process in a way that is not concededly conventional and/or trivial) or 2) transforms an article from one thing or state to another. Since software has to run on something, as a practical matter, this is a difference in semantics rather than substance for most software patent applications.

Although software patents are now common—in fact, they've become more and more integral to the survival and growth of large and small software companies—many software developers continue to rely on copyright protection, which protects their code from being plagiarized. Copyright protection is free and automatic; it attaches to a program the minute it is written, whether or not the author registers it with the copyright office. In the past decade, courts have allowed more of the functional aspects of software programs to be protected by copyright, but the courts can go only so far in extending a doctrine meant to protect creative expression rather than functional ideas.

In contrast, a patent prevents *reverse engineering*, in which competitors analyze the software product to develop their own functionally equivalent programs.

A software patent holder has a tremendous competitive advantage under the *doctrine of equivalents* (see page 191).

Patenting Software

Software enhancements and innovations that address a specific, previously unsolved problem often have the requisite level of novelty to qualify for patent protection. If you are a software developer, or if your invention involves software, here are a few reasons you might want to consider pursuing patent protection.

1. Patent protection is a broader form of protection. Copyright protection is helpful in situations of outright piracy, but it is not well suited to deter an aggressive competitor. Copyright does not prevent your competitors from creating functionally equivalent code through reverse engineering, black box testing, or other methods.

Patents, on the other hand, carry more valuable and expansive rights, conferring a monopoly type of protection from competitors trying to encroach on the market you've established.

A well-drafted patent can give you a monopoly with respect to the innovative, functional elements of your software. It allows you to prohibit a competitor from marketing a program that performs the same innovative tasks as your patented product, even if that competitor has not copied any of your code.

With a patent, there's no need to prove that a third party has actually copied key elements of your work; only that they've infringed your patent by selling a program that falls within the scope of what you've claimed in your patent.

2. Most new software contains some protectable elements. Do you feel that there is a market for your software because it fills a need for increased efficiency or offers a new and useful solution to a problem? If so, it's likely that at least some aspects of your software may be considered novel enough to patent.

Section 101 of the U.S. Patent Act states:

> *Whoever invents or discovers any new and useful process, machine, manufacture, or composition of matter, or any new and useful improvements thereof, may obtain a patent, subject to the conditions and requirements of this title.*

The Amazon.com "one-click" technology was patentable because it contained novel features in addition to relying on previous e-commerce technology. Software patents cover a range of algorithms and methods for performing calculations and processing numerical and financial data. Examples include insurance application processing, loan application processing, stock/bond trading and management, health-care information management, reservation systems, auction systems, and so on. Your attorney can help you identify the elements of your software that are sufficiently novel to warrant patent protection.

3. Provisional applications can make patent protection economical for small software developers and startup businesses. A provisional application is a simplified patent application and is less costly to file than a standard application. Consider filing a provisional application in order to establish the earliest possible filing date for your patent while you continue to perfect and establish a market for your new software.

In a provisional application, you must sufficiently identify your invention, but your attorney need not draft the detailed claims defining the parameters of your invention that are required in a standard application. Filing a provisional application entitles you to use the "patent pending" designation on your software. On or before the expiration of one year, you must file a standard application with the usual detailed claims (which will, if approved, ultimately give you protection twenty years from filing). Your patent is given the earlier filing date of your provisional application.

Simply put, you have one year from the date that you disclose your invention or offer your invention for sale. Within that one-year period, you can file for a provisional application which, in essence, gives you up to one additional year in which to file a nonprovisional application.

Most importantly, don't be overwhelmed. No one expects you to do this all by yourself. See page 113 to

Ten Tips for Independent Inventors

1. View a patent as a tool to protect something that is marketable and therefore potentially valuable, not a piece of paper that's going to make you rich.

2. If you know you are going to file for patent protection, do so sooner rather than later to protect your rights.

3. Use provisional patent applications before you have tested the market and fully developed your invention.

4. View a patent as a legal document that defines important rights; don't file one without consulting a registered patent attorney.

5. If you begin marketing your invention and later make changes that will add value in the marketplace, amend your pending application or file additional patents to protect those innovations.

6. Visit the USPTO website at www .uspto.gov and become familiar with the free resources that are available for individual inventors.

7. Read your patent application before your attorney files it. You, as the inventor, should fully understand every word in the claims and be able to assist your attorney in identifying possible additional embodiments of your invention that should be covered in your claims.

8. If anyone has contributed to the development of your invention, he or she must be named as a coinventor. Ask your attorney about having any coinventors assign their rights to you before filing the patent application.

9. Stay involved in the patent prosecution. In the event of an "office action" with a USPTO examiner, you have the opportunity to talk in person with the examiner to help structure the allowable claims.

10. Ask your lawyer for a checklist of things you need to do after your patent issues, such as paying maintenance fees to the USPTO at 3½, 7½ and 11½ years. Abandoned patents open the door to hungry competitors.

read more about provisional patents and consulting a patent attorney to help protect your idea, and visit the USPTO's website at www.uspto.gov for more information.

4. Patents may be valuable portfolio assets that can attract investors. Investment bankers routinely analyze the patents held by a company as a key indicator of its net worth, and they are therefore familiar with the vernacular of patents. There can be a correlation between the strength of a patent portfolio and the financial value of a company's stock when the market for the product is strong.

Ease of enforceability is another factor that makes a patent a more attractive asset than a copyright to a potential

(continued on page 110)

INVENTOR PROFILE

Jerome Lemelson: the Most Controversial Inventor

Jerome Lemelson ranks just behind Thomas Alva Edison as the most prolific patent holder of all time. Lemelson, who was born in 1923 and educated as an engineer, acquired more than 550 patents during his life, approximately one patent a month during his forty-five-year career. Critics say that Lemelson built a career on shrewdly predicting the trend of emerging technology and acquiring a patent directly in its path. He amassed millions of dollars asserting, litigating, and settling patent disputes.

He obtained patents on a dazzling array of technologies and products that seemed to encompass every aspect of American life: crying baby dolls, bar-code readers, cordless telephones, cassette players, camcorders, fax machines, robotics, personal computers and peripherals, and manufacturing processes spanning dozens of industries in which no one had ever known him to be personally involved. He managed to do all this without the backing of major research institutions or big corporations.

Eventually, in the later years of his career, Lemelson spent the majority of his time defending his patents in court. In 1998 he began sending letters to customers of two companies that manufactured and sold bar-code scanners and related products, accusing the manufacturers

Fax machines and other communication devices represent only a fraction of Lemelson's patents.

of patent infringement. Both companies sued to have Lemelson's bar-code-related patents declared invalid. The patents in question were initially filed in the 1950s and were pending for decades. Lemelson had legally manipulated the patent system so that the applications had been pending, but were not prosecuted, for a period of more than thirty years. On appeal, the court declared Lemelson's patents unenforceable on the basis of unreasonable delay.

Lemelson became most famous for perfecting a legal maneuver called the "submarine patent." Using this strategy, he positioned himself to extract large settlements from hardworking inventors by rushing to file patent applications for technologies such as the bar-code scanner or the crying baby doll. Lemelson would keep his pending patent applications secret (like a submarine under water) until the inventions were commercially successful. He would then produce his patent and threaten to sue for infringement, negotiating large amounts of money to settle the case.

In 1996 Lemelson was diagnosed with liver cancer. He submitted nearly forty patent applications during the last year of his life, including several for improved medical devices and cancer treatments. Prior to his death, he established the Lemelson Foundation, which has donated more than $50 million to helping individual inventors.

Lemelson's strategies inspired Congress to pass laws requiring inventors to publish their patent applications within 18 months of filing to discourage Lemelson copycats.

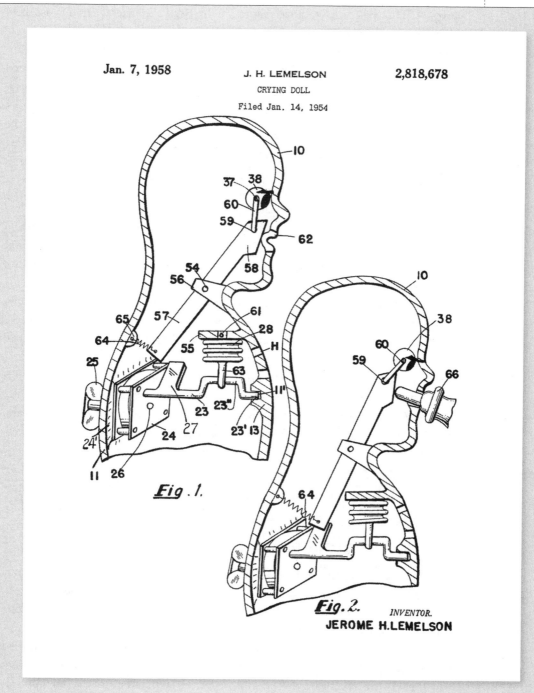

Jan. 7, 1958   J. H. LEMELSON   2,818,678
                          CRYING DOLL
                      Filed Jan. 14, 1954

Fig.1.

Fig.2.   INVENTOR.
JEROME H.LEMELSON

Jerome Lemelson's "crying doll" patent was achieved through a legal maneuver now called the "submarine patent."

(continued from page 107)

investor. U.S. patent law carries more than 200 years of court precedent. In contrast, the "look and feel" doctrine of copyright is relatively new and is a comparatively nebulous concept for judges to apply.

5. Patents can bring greater certainty to licensing arrangements. Licensing is an explosive trend. IBM, one of the first of the U.S. companies to adopt an aggressive patent licensing strategy, saw its annual patent licensing revenue increase from about $30 million in 1990 to well over $1 billion by 2001.

Licensing allows software developers to simultaneously receive ongoing revenue from numerous users. However, before licensees will pay a substantial fee, they generally want to know if you hold a valid patent or other intellectual property rights to the software you're offering. They want assurance that they won't be sued for infringing someone else's patent.

You as the inventor must demonstrate diligence in your process, including prompt filing of a patent.

---

6. Patent protection is available for software algorithms. Patenting software inventions has become easier since the State Street Bank decision of 1998, which held that a software invention is patentable if it produces a "useful, concrete, and tangible result."

In 1999 the federal circuit also held that software inventions that include algorithms are patentable under limited circumstances. The court stated in *AT&T Corp. v. Excel Communications, Inc.* that if a software invention includes a mathematical algorithm, and if the mathematical algorithm is "applied in a practical manner to produce a useful result," then the invention is patentable.

7. Patents protect you against a growing number of international competitors. Many countries do not extend patent protection for software, but the list of the ones that do is growing. Currently, Australia, Canada, England, Germany, Korea, and Japan have laws for protecting software.

The European Patent Office extends patent protection to software that improves the functionality of hardware, facilitates the interoperability of software, or creates a new "technical effect." You have the option of filing a European patent application (which covers most European countries), rather than filing individually in each European country.

8. Delay in filing may result in loss of your rights. A delay in filing may permanently jeopardize your ability to obtain patent protection for your software innovation.

Generally, U.S. patents are awarded based on who is first to invent, rather than who is first to file an application. In the event of a dispute, you as inventor must demonstrate diligence in your process, including the prompt filing of a patent. In addition, you risk your right to obtain patent protection if you disclose any part of the invention prior to the application date to any third parties who are not under a duty of confidentiality. Recently the federal circuit court denied patent protection to an inventor who had explained his software invention to a coworker in enough detail to allow the coworker to "practice" it (even though the software used to practice the invention did not exist at the time it was discussed).

If you intend to market your invention, consult your attorney about creating a nondisclosure agreement as well as filing a provisional patent application that will preserve your patent rights and deter illegal copying or imitation of your product or process.

Biotechnology Patents: The Mouse That Roared (Legally)

Since Dolly the sheep was cloned, no one in the scientific community has doubted that human cloning would soon follow. Most countries in Europe have already adopted legislation denying patents for processes that involve human cloning.

Biotechnology patents are a favorite of the tabloids and provide fodder for political and ethical debate. Biotechnology uses advancements, such as DNA technologies, to create processes that alter biological systems and life-forms.

Dolly the sheep, though controversial, inspired legislation prohibiting human cloning patents.

In 1980, amid considerable controversy and on the heels of a decision protecting patent rights for genetically altered bacteria, the U.S. Supreme Court awarded Harvard University a patent for a genetically altered mouse (see next page). The mouse was specially developed to carry cancer cells in its genes and develop tumors quickly. The "Harvard mouse patent" was significant because it was the first patent to be granted for a new animal life-form.

More recently, stem cell patents have made headlines. University of Wisconsin researcher James Thompson was the first to isolate embryonic stem cells, a discovery in 1988 that was later the subject of three issued patents. Advocates of embryonic stem cell research argue that these patents offer the best hope for potential cures for conditions such as diabetes, Parkinson's disease, and spinal-cord injuries. The

United States Patent [19]

Leder et al.

[11] Patent Number: **4,736,866**

[45] Date of Patent: **Apr. 12, 1988**

[54] **TRANSGENIC NON-HUMAN MAMMALS**

[75] Inventors: **Philip Leder, Chestnut Hill, Mass.; Timothy A. Stewart, San Francisco, Calif.**

[73] Assignee: **President and Fellows of Harvard College, Cambridge, Mass.**

[21] Appl. No.: **623,774**

[22] Filed: **Jun. 22, 1984**

[51] Int. Cl.⁴ C12N 1/00; C12Q 1/68; C12N 15/00; C12N 5/00

[52] U.S. Cl. **800/1;** 435/6; 435/172.3; 435/240.1; 435/240.2; 435/320; 435/317.1; 935/32; 935/59; 935/70; 935/76; 935/111

[58] Field of Search 435/6, 172.3, 240, 317, 435/320, 240.1, 240.2; 935/70, 76, 59, 111, 32; 800/1

[56] **References Cited**

U.S. PATENT DOCUMENTS

4,535,058 8/1985 Weinberg et al. 435/91
4,579,821 4/1986 Palmiter et al. 435/240

OTHER PUBLICATIONS

Ucker et al, Cell 27:257–266, Dec. 1981.
Ellis et al, Nature 292:506–511, Aug. 1981.
Goldfarb et al, Nature 296:404–409, Apr. 1981.
Huang et al, Cell 27:245–255, Dec. 1981.

Blair et al, Science 212:941–943, 1981.
Der et al, Proc. Natl. Acad. Sci. USA 79:3637–3640, Jun. 1982.
Shih et al, Cell 29:161–169, 1982.
Gorman et al, Proc. Natl. Acad. Sci. USA 79:6777–6781, Nov. 1982.
Schwab et al, EPA–600/9–82–013, Sym: Carcinogen, Polynucl. Aromat. Hydrocarbons Mar. Environ., 212–32 (1982).
Wagner et al. (1981) Proc. Natl. Acad. Sci USA 78, 5016–5020.
Stewart et al. (1982) Science 217, 1046–8.
Costantini et al. (1981) Nature 294, 92–94.
Lacy et al. (1983) Cell 34, 343–358.
McKnight et al. (1983) Cell 34, 335.
Binster et al. (1983) Nature 306, 332–336.
Palmiter et al. (1982) Nature 300, 611–615.
Palmiter et al. (1983) Science 222, 814.
Palmiter et al. (1982) Cell 29, 701–710.

Primary Examiner—Alvin E. Tanenholtz
Attorney, Agent, or Firm—Paul T. Clark

[57] **ABSTRACT**

A transgenic non-human eukaryotic animal whose germ cells and somatic cells contain an activated oncogene sequence introduced into the animal, or an ancestor of the animal, at an embryonic stage.

12 Claims, 2 Drawing Sheets

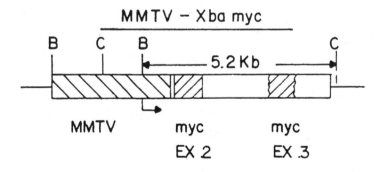

The "Harvard mouse patent" was the first patent granted for a new animal life-form.

cells, which are taken from days-old embryos, work as a type of master cell that is capable of changing into many kinds of tissues and cells. Opponents of stem cell research call it immoral because it requires the destruction of the embryo. In May 2007 the USPTO opted to invalidate Thompson's patents during a process called a reexamination, on the grounds that the inventions were "obvious to one of ordinary skill."

Provisional Patent Applications: A Smart Strategy for the Cost-Conscious Inventor

Provisional patent applications are one of the best bargains available for individual inventors who need to protect their inventions quickly and have a limited legal budget. However, a great deal of misunderstanding surrounds them.

For starters, there is no such thing as a "provisional patent." A provisional patent *application* is a detailed description of your invention that contains all the information that will form the basis of the claims in your patent. However, the provisional application need not contain any actual claims.

Provisional applications are documents that you file with the USPTO before you're ready to file a regular utility-patent application. A provisional application will *never* be examined by a patent examiner, but will be kept on file for a one-year period. By the end of that period you must file a regular (non-provisional) patent application, which contains specific and detailed claim language, and which will be examined by the USPTO to determine the patentability of your claims.

Since the provisional application contains no claims and isn't examined, it is a lot less costly than applying for a patent. It gives you a year to test the market and refine your invention before having to pay a lawyer to draft detailed legal claim language. This could be a very useful way to test the market and determine whether the investment in moving forward is likely to yield a meaningful return. It also gives you the ability to "shop the idea around" to prospective licensees while having legal protection. It establishes your filing date and protects your rights (both domestic and foreign) in the event that you make an enabling disclosure. It is a much stronger type of protection for your patent rights than a nondisclosure agreement is. An even bigger advantage is that once the provisional patent application is filed, you are allowed to use the "patent pending" designation on your invention to warn the competition.

The downside of a provisional patent application is that it delays the start of the patent examination process for up to a year. You will also lose the benefit of your provisional application if you don't file the nonprovisional application within one year of the date on which you filed the provisional application. There are no extensions and no exceptions to this rule.

The Battle over Business Method Patents

Business method patents are close cousins to software patents. Both can trace their modern origins to a 1996 case of great legal significance. In the famous *State Street Bank v. Signature Financial Group* case, the Court of Appeals for the Federal Circuit upheld a disputed patent involving a software system for tracking the performance of pooled mutual fund investments for partnerships. The court held that mathematical algorithms (namely, software programs) could be patented if they were applied to produce a "useful, concrete, and tangible result."

Since the court of appeals decided the State Street Bank case, the USPTO has issued hundreds of business method patents, and thousands more are pending.

Interestingly, the U.S. Supreme Court has never taken a look at the specific standards that should apply to business method patents. (To date, the existing standards are based on case results at the circuit court level, but the Supreme Court has recently hinted it will take a hard look at them in the near future.)

Trademarking Your Product

W e've talked a great deal about protecting your invention by applying for a patent; there are other ways to protect your ideas and product as well.

In many cases, a *brand* can become a very effective way to distinguish your product from that of your competitors. Being the first to enter the market or having an innovative product is important, and even more so if you can distinguish your product with a name and identity that are exclusively yours. In short, crossing the finish line first can be even more lucrative if your product looks good doing it. This is accomplished when you file for a trademark to protect the name and identity of your product. American businesses, from fledgling startups to vast multinational corporations, spend billions of dollars every year to develop and protect their trademarks. Inability to protect your marks can cost your company its competitive edge, which can mean that an otherwise superior product or service will fail in the marketplace.

FROM THE ENTREPRENEUR
I'VE GOT MY BRAND ON YOU

A brand is just as important as the product itself—and is just as necessary to protect. Think about the most successful products in America: Coca-Cola, the Big Mac, Google. Not only do these names recall the flavor, taste, or function of the product itself, but we see the products' brands in our minds as well. It's practically impossible to think of a Big Mac without seeing the golden arches. Trademarking is vital to the invention business because it protects your product's style—which is just as important in the marketplace as protecting its function.

Trademarks can be an enormously valuable asset for a company, the value of which is measured as *brand equity*. Take, for example, the beverage Snapple. The company was created in 1972 by three partners who sold fruit drinks to health food stores in New York City. The brand name was short for Snappy Apple, a carbonated apple-juice product whose brand they purchased from a Texas man for $500. As the company grew, so did brand loyalty, and by 1994 sales approached $700 million. The following year, Quaker purchased the company for $1.7 billion. If you added up the inventory, equipment, and all other assets, the sum was far less than the purchase price. The difference was the value of the brand. The purchaser, Quaker, was buying the name recognition and goodwill associated with it.

It is also important to note, however, that brands can lose value when they are not managed properly. Following the sale of Snapple, other brands entered the market with better products, packaging, and marketing. As a result, Snapple lost much of its appeal, as evidenced by the 1999 sale to TriArc for only $300 million! Yes, that's correct. In less than two years, the value of the brand decreased by more than $1.4 billion.

What Constitutes a Valid Trademark?

Trademarks protect the words and symbols that identify a product, not the product itself. Any word, symbol, slogan, logo, device, or design that uniquely identifies a product can be legally protected as a trademark. The sole legal purpose of any trademark is to identify the source of goods and services. Unlike other types of intellectual property, such as patents and copyrights, trademarks bear no relationship to invention or discovery. In fact, trademarks must *not* have a function other than identifying a product. If they do, their owners must generally protect the mark as a component of a product's design under patent or copyright laws.

Establishing Legal Rights Within Your Trademark

Getting legal rights to a trademark is not as complicated as some lawyers might want you to think. To get federal trademark rights, you must actually use the mark in some way that associates it with your product. Although registration of a mark is important, it is really the first use of the mark that establishes the legal rights in the trademark.

Until fairly recently, U.S. businesses faced a miserable dilemma. Federal law wouldn't allow them to register a trademark unless they could prove their prior use of it. This meant they had to spend money promoting a mark and risk associating it with another emerging product or service line without the benefit of registration. In 1988 federal trademark law was revised to allow applicants to file "intent to use" applications on trademarks. Once one is filed, you must use the mark within six

The value of the Snapple brand in 1995 was estimated at $1.7 billion.

months to make it valid. Extensions for additional six-month periods may be filed, but this can be costly.

Protecting the "Trade Dress" of Your Mark

One type of trademark asset inventors and businesses often don't realize they have is *trade dress*. Trade dress is the total appearance of a product or service. Although you can register trade dress with the USPTO or file for design patent protection, it's not always necessary. You acquire trade dress just by using it in the marketplace and being able to demonstrate that the public associates the appearance of your product with its source. There are many court cases in which one party has been found to be misappropriating another party's established trade dress, or in which the first party is found to be using a product or package design that obliquely references that of another product (owned by the second party) that recognizably uses that form or mark.

Marks that are easy to defend in the face of misuse or infringement by a competitor are known as *strong marks*.

This type of trade dress is called product configuration. If the shape or appearance of a product has no function other than identifying the source of the product, it can be protected as trade dress. Examples of protectable product configurations are the plastic ReaLemon brand lemon-juice container, which looks like a lemon, or the bottle for Mrs. Butterworth's syrup, which is shaped and decorated to look like a plump woman in an apron.

Developing a Strong Mark versus a Weak Mark: Types of Protections

Looking to save on future legal fees? Then it's important to consider choosing a "strong" trademark rather than a "weak" one.

Some trademarks are legally stronger and therefore easier than others to protect against infringement. Marks that are easy to defend in the face of misuse or infringement by a competitor are known as *strong marks*. It's generally true that the less descriptive your mark is of the actual product, the stronger your case will be against infringers. Trademarks are placed into four general categories, based on their relative legal strength: Fanciful marks (for instance, Google, which is completely made up) are the hardest for third parties to challenge, making them legally strongest. Subjectively chosen marks (Apple) are only slightly less strong. And suggestive marks, which suggest the characteristics of the product being offered (Jiffy Pop), are the hardest to protect. Purely descriptive marks (such as flying disc) can't be protected at all, since they'd prevent people from using necessary language. If you choose a mark that is legally weaker, you won't be able to exclude other users of the trademarked terms as broadly. You'll have to prove that alleged infringers have a mark that is a lot more similar in look, sound, or meaning than you'd have to in fighting infringement of a stronger mark. Also, where products are not exactly similar but are closely related, the owner of the stronger mark will be able to exclude similar marks, whereas the owner of a weaker mark will not be able to do so.

Businesses such as Google and Microsoft opted for strong marks. They made up their own words, which were easy to register with the trademark office and to protect from competitors. These companies relied on the strength of advertising and marketing to build their brands and give these otherwise meaningless terms extraordinary value in the marketplace. Other companies opt to use more descriptive marks (like NutraSweet or Handi Wipes),

knowing that they are legally weaker and harder to protect. These companies take the view that there is an advantage to a more descriptive mark because it has *market strength*: it educates, and is absorbed by, the public more quickly.

Death by "Genericide"

What do aspirin, baby oil, cellophane, and shredded wheat all have in common? They're all perfectly good, distinctive trademarks that have become "genericized" over time.

A trademark becomes a generic term when a court or the United States Patent and Trademark Office finds that, in the mind of the public, the mark has come to represent particular goods or services rather than to describe the origin of those goods and services (person or company). Though it's a little counterintuitive, this spells disaster for a company. All of its prior advertising dollars and marketing efforts are lost.

OH SO GENERIC

FORMER TRADEMARKS	PRODUCT
baby oil	mineral oil
aspirin	acetyl salicylic acid
escalator	moving stairways
hoagie	"submarine" sandwich
honey baked ham	ham with sweet glaze
Murphy bed	type of bed that folds into wall
shredded wheat	baked wheat cereal
thermos	insulated bottle
trampoline	athletic apparatus for jumping

Making Their Mark

See how many brand names you recognize simply by looking at their mark. A successful mark is clear and specific but remarkably versatile.

1-NBC, 2-Woolmark, 3-Bell Atlantic, 4-Microsoft, 5-Chevrolet, 6-CBS, 7-Chili's, 8-Nike, 9-Playboy, 10-Target, 11-Warner Bros., 12-Apple.

In order not to become a victim of your own advertising campaign, as the marks in the table below did, take precautionary measures as to how your trademark is actually used. Jell-O, Band-Aid, and Google all provide excellent role models in this regard. Google sends letters to the media requesting that they not use the term "google" as a verb and instead use "Google" as an adverb, Google search, to avoid having their mark become genericized. While you might think a company would be happy that consumers are using its product name so frequently that it becomes synonymous with, say, a word as common as "search," what it means to the company is that the consuming public has stopped recognizing the specific brand name that the company has worked so tirelessly to individualize. The genericized product essentially becomes no different from the field of competitors—the brand itself ceases to exist. Jell-O calls itself "Jell-O brand" gelatin for this same reason—watch for this the next time you see a commercial. Johnson & Johnson now promotes "Band-Aid brand bandages." They use the generic term "bandages" in order to protect their trademark and prevent other brands of adhesive bandages from benefiting from broad consumer identification with Band-Aid. Kleenex avoids genericizing its mark by making sure that its brand name for a tissue is always capitalized and that its advertising materials make clear that it's a registered trademark for disposable paper products.

Trademarks visibly represent your product in the marketplace, and need to be protected as fiercely as patents.

Maintaining Your Mark

To avoid having your mark become "genericized," there are some cardinal rules of trademark usage.

■ **Use the** ™ **next to the mark as soon as you begin the trademark process.** This alerts the public of your intention to protect the mark.

■ **Use the** ® **symbol to denote a federally registered trademark.** This symbol can be used only after you receive a federally registered trademark, and should be used in any materials promoting goods and services covered by the mark.

■ **Use the mark as an adjective.** Don't use the mark as a noun or a verb, but rather to describe the source of the goods. ("I'm going to use the Xerox copier" instead of "I'm going to Xerox this.")

■ **Don't use the mark as the object of a plural or possessive term.** You won't find Johnson & Johnson allowing a retailer to say, "We sell a lot of Band-Aids." Instead, you'll hear and see references to a "Band-Aid brand bandage."

■ **Establish company policies and guidelines.** Establish and enforce clear policies as to how suppliers, distributors, retailers, and promoters should use your mark by directing them not to use it as a verb, as a plural, or as the object of a possessive.

■ **Use your mark consistently.** Make sure your mark is spelled and capitalized consistently and correctly and that correct punctuation, such as a hyphen, is used. For example, Coca-Cola should never appear as Coca cola; and "Kids prefer Jell-O" should never appear as "Kids prefer jell-o."

As an inventor, you have an impressive arsenal of intellectual property assets that you can assemble to protect your invention. Patents, trademarks, copyrights, and trade secrets all serve to create barriers to entry and, more important, act as appreciable assets that could be worth a significant amount if they are properly developed and protected.

Johnson & Johnson will have you asking for a "Band-Aid brand bandage."

CHAPTER 5

Make Your Mark

Branding, Pitching & Selling

Coca-Cola, perhaps the best-recognized brand of soda in the world, was invented in 1885 by John Stith Pemberton, a pharmacist in Atlanta, Georgia. Pemberton originally dubbed his invention Cocawine and sold it as a medicinal tonic at the soda counter of Jacobs Pharmacy for five cents a glass. It contained a stimulant made from coca leaves from South America (the plant from which the drug cocaine is derived) and was flavored with caramel syrup and high-in-caffeine cola nuts.

In 1886, when the city of Atlanta passed a regulation forbidding the sale of wine, Cocawine went through its first

Coca-Cola is one of the strongest and most recognized brands in the world.

rebranding (even though it was nonalcoholic, being associated with an illegal beverage wasn't ideal). Pemberton began promoting the drink as a carbonated, nonalcoholic version of a French wine product. He launched an ad in the *Atlanta Journal,* which boosted his soda-counter sales to 300 drinks a month. Pemberton's bookkeeper came up with the name Coca-Cola and wrote it out in the distinctive handwriting that appears on cans and bottles and across billboards worldwide today.

That same year, Pemberton started selling off his business interests in Coca-Cola. By 1888 three versions of the soda, sold by three separate, competing companies, were on the market. Asa Candler, the original buyer, attempted to sell the beverage under the brand names Yum Yum and Koke. Even though the formulas were identical to Coca-Cola's, both brands were a failure. Consequently, Candler convinced Pemberton, shortly before Pemberton's death, to sell him "exclusive" rights to the Coca-Cola name. Those rights are held by the company to this day. And Candler's branding of Coca-Cola, which began when he launched a large-scale advertising campaign offering coupons for free tastes of the refreshment and supplying pharmacies with products emblazoned with the Coca-Cola logo, were the first steps on the product's road to widespread fame.

The Coca-Cola Company has continued to grow and has one of the most recognized and admired brands in the world. In addition to the flagship products (Coke, Diet Coke, Sprite, and so on), the company has created or acquired dozens of other brands such as Minute Maid, Glacéau Vitamin Water, and Honest Tea.

Branding and Trademarks: Turning an Invention into a Product

Branding creates an identity for your invention and stands as your promise of quality and consistency to the consumer. The term refers to the collective ideas and concepts associated with your product that cause consumers to buy it—often without fully realizing why. A powerful brand is like an invisible magnet that draws a consumer to choose one product over another.

How Branding Works on the Brain

Have you ever passed a billboard with a product name or a single image that didn't seem to "say a lot"? Maybe you passed the same ad later somewhere else and still didn't think too much about it. What you probably didn't realize was that your brain was being deliberately affected by a branding image each time you came across it. Our brains process product brand images without our conscious knowledge. When we are confronted with a decision between two brands, we tend to respond well to the more familiar one because our brains process it more easily as easily accessible "shorthand data."

An interdisciplinary team of researchers at a meeting of the Radiological Society of North America in Chicago presented a 2006 neurological study that measured how much brain energy and activity were necessary to respond to a brand. The study concluded that the brain has to work harder to process images and information associated with "weak" brands than with stronger, more easily identifiable ones. Advertisers have long known this.

By some estimates, the average consumer sees more than 3,000 brand messages a day and spends more than three years of his or her life watching television commercials. By the time you've brushed your teeth and taken a shower in the morning, you'll have seen perhaps a dozen brands: printed on your toothpaste tube, imprinted on your soap bar, engraved in your bathroom plumbing parts, and embroidered onto the labels on your clothes. If you had to process this information consciously, well, you simply wouldn't have time to! Thankfully, the human brain is very sophisticated. It stores and accesses images associated with the brands (as it does with any other familiar information, like your telephone number or home address) without voluntarily calling them up. For example, when you see the label on your toothpaste, you may be able to recall an image of the white-toothed model who was hawking it on TV that morning; the distinctive smell of your soap may bring up the image of the packaging it came in.

In a world filled with millions of products competing for billions of dollars, a brand that demonstrates that it can draw consumers is a valuable economic asset that can actually be reflected on a company's financial statements in the form of trademarks and design patents. A brand is something that can be protected legally by trademark law, and that can be sold to others exclusively or nonexclusively (simultaneously). You can also license others' right to use your brand in connection with their products if you so choose. It takes time and a well-executed strategy to develop a brand identity for your invention (product), and there is more than one way to do it. For example, as with Befudiom (see sidebar, next page), a team of marketing experts can give a product a strong brand identity by licensing a well-known brand name.

Branding Your Product So People Will Buy It

There may be plenty of logical reasons for consumers to buy your product, but most decisions to purchase are made instantly, often at the point of purchase and without much analysis of facts. Branding provides a better impetus for customers to pick your product without going through a conscious decision-making process.

Sparkling, toothy smiles are associated with brands of toothpaste.

CASE STUDY
Branding for Recognition

Inventor Wendy Hampton of Lawrenceville, Georgia, never intended to develop the next big party game. This single mom just wanted to create something educational and fun that she could play with her ten-year-old daughter.

When her daughter came to her and proclaimed that she was "totally bored," the two sat down on the floor and wound up with an idea for a game about idioms, those quirky phrases and slang terms that we use daily but whose meanings aren't always apparent. Soon Wendy was staying up late at night—every night—developing content and gathering thousands of idioms, such as "kick the bucket," "axe to grind," or "holy cow."

Befudiom has the appeal of many of America's favorite party games; it invites players of all ages to compete to uncover the meanings behind those "befuddling" phrases by acting, drawing, shouting, or spelling. Fun and educational, Wendy's invention sailed through the *Everyday Edisons* casting call. Everyone who saw it loved it.

The game market is a tough one to crack, though, being dominated by established titans like Milton Bradley and Hasbro. The *Everyday Edisons* team decided that Befudiom needed a branding boost. Despite having a strong name that evoked perplexing fun and puzzlement, the marketers felt that licensing an existing brand would add credibility. Merriam-Webster came with a strong word-based track record, having published the first American English dictionary. It was well-known and recognized by consumers, making it a perfect partnership for the game. After a brief negotiation, it was decided to license the Merriam-Webster brand and call Wendy Hampton's game Merriam-Webster's Befudiom. With a strong brand, the product has a better chance of being chosen by consumers based on its brand recognition.

Partnering with Merriam-Webster gave Befudiom a brand boost.

It's up to you to decide what unique qualities are important about your product and your brand. For example, Miller Brewing associates the feeling someone has at the end of a workday with their slogan "It's Miller Time."

A product announcing "Just Do It" communicates Nike's association with personal competitiveness.

Even if you're a small company or an individual inventor with a single new product, buyers at major retail chains who review your product will place as much weight on the *image* of value it conveys as they will on the actual price and value of the product. Although you are not touting an established brand, your invention will have its own unique image, whether you consciously create it or not. Will consumers skimming store shelves view your product as a good value and associate it with qualities they want to possess? Will its color, design, and packaging convey its function, and will the product image be reinforced (and perhaps repurchased) upon return trips to the store? Product name, packaging, and image-laden slogans can strongly influence a major retailer's decision to carry your product, as well as a consumer's choice to buy it.

What instantly springs to mind when customers hear your product's name and see its packaging? If you're uncertain, it's time to put as much thought into the branding of your invention as you did into developing it. This doesn't have to cost millions or take years to accomplish. Here are some guidelines for creating a winning brand image:

Differentiate your brand. No one product can be all things to all people, so you must ask yourself what function or niche your product fills. How is it different from products people have purchased in the past and the competing products that they might purchase in the future? What does your product or service deliver that's valued most by

its users? Does it save time or money, or allow them increased comfort? Try jotting down a few words that convey your product's most important differentiating qualities. Remember, branding is all about sending a strong and consistent message.

Wii branding is playful to reflect its fun technology.

Select a product name. Your brand or product name should either evoke images you want people to associate with your product (Nike, named for the Greek goddess of victory) or be a whimsical, unrelated word (Wii, a made-up word easily pronounced internationally) with which you can suggest future value. The North Face is a name which consumers have learned to associate with outdoor clothing. Upon seeing a CoverGirl cosmetics product, the slogan "easy, breezy, beautiful" comes to mind with an image of a beautiful smiling woman. And thanks to Nike, we all know now to "just do it." (See pages 115–119 for more discussions of legal and trademark considerations when choosing a name.)

Begin associating words, phrases, images, symbols, and logos with your product. A brand is really a collection of impressions that potential customers and the public have about your product, which, in turn, increases its value over time. What immediately comes to mind when you see "golden arches"? What about a dark brown delivery van? Or a silhouette of an apple? These are examples of how logos, colors, phrases, or images are used to differentiate manufacturers' products or services.

CASE STUDY

What's in a Name? Frisbees and Other Identified Flying Objects

"Frisbee." It's a word that instantly conjures up a vivid mental image. But the "Pluto Platter?" Mention it to your friends at a picnic or backyard gathering and you're likely to get blank stares in return. Fred Morrison, inventor of the flying plastic disc, dubbed it a Pluto Platter in the hope of capitalizing on the American public's growing fascination with UFOs and flying saucers. But the name never really "took off," so when Wham-O cofounders Rich Knerr and Arthur "Spud" Mellin obtained the marketing rights to Morrison's invention in 1957, they quickly changed its name from Pluto Platter to Frisbee, after learning of a pie-plate-throwing craze at New England college campuses. (The word "Frisbie" was stamped on the tin plates of pies sold by the popular Frisbie Pie Company of Connecticut, and students would yell "Frisbie!" before throwing, just like golfers yelling "Fore!" Knerr and Mellin changed the spelling of the word to "Frisbee" to avoid trademark infringement.) Morrison wasn't convinced it was a good move—in fact, he thought it was a terrible idea. However, Frisbee sales quickly soared, and the company went on to sell hundreds of millions of Frisbees. Today, the Frisbee name is owned by Mattel, Inc., and is recognized all around the world.

Packaging Your Product

Every time customers pass your product on a store shelf, they have a critical branding experience. When people see your product for the very first time, it's the packaging and presentation that they see, not the product itself. Packaging has a big impact on their willingness to invest in trying something new. Even before it gets to store shelves, packaging also has particular influences over the decisions of major retailers and distributors to carry your invention. Usually, the more polished the packaging and presentation, the better. Impressive packaging can help you elbow into a spot in a crowded marketplace and give your potential customers a critical branding experience every time they encounter it on a store shelf.

There are two cost components to the packaging equation. The first is the graphic design, and the second is the structural appearance. Working with a professional packaging-design firm is the preferred way of creating packaging if budget is not an issue. Price can vary from company to company and product to product, but $5,000 to $50,000 is a very realistic expenditure range for obtaining professional results. There is a reason these firms charge what they do for their services: A great deal of work goes into designing effective packaging. The goal of good packaging is to present a product that gets noticed and picked up off the shelf. If this happens, the likelihood of a purchase increases significantly.

Alternatively, you, as the inventor, may feel that you can design the packaging for the product on your own. Though this is not generally recommended, it can be effective if you have the necessary skills. Make sure you get feedback before taking what you've designed to a retailer. You may find that your packaging lacks the professional appearance that retailers demand today.

In determining your initial manufacturing costs, you should be sure to allocate at least 15 to 20 percent for packaging. The cost can vary widely. Perfume, for example, has significant packaging

costs in comparison with the actual price of the ingredients. Tools, on the other hand, may be "packaged" with just a sticker, which costs practically nothing. It's worthwhile to overestimate the cost of the packaging you might require at the start so you can avoid skimping in the end. Your manufacturer can generally produce your packaging or refer you to someone who will. Graphic-design services may or may not be included, but remember to inquire about them and budget accordingly.

Your packaging should quickly convey the name and function of your product; this is particularly true if you have a product that fulfills a novel function. For example, if you are selling a device that makes heart-shaped

CASE STUDY
Naming the BlackBerry

The BlackBerry, introduced in 1997, is a wireless handheld device that supports an e-mail system that has been lauded as the most efficient on the market. (It is called the "push e-mail" system because it downloads e-mail to the user's inbox instead of requiring the click of a button to send and receive.) It offers mobile telephone, text messaging, i-faxing, Web browsing, and other wireless services, but with a focus on e-mail. Being positioned primarily as a business tool, BlackBerry has edged out competition such as the Palm Treo and has maintained its market despite the popularity of the iPhone.

Early on, the name Leapfrog, alluding to the giant leaps forward in technology, had been considered. After this, the manufacturers of the BlackBerry (Research In Motion, Limited) rejected the name Strawberry after a product-naming expert suggested that the miniature buttons resembled the tiny seeds of a strawberry (a linguist at the firm thought "straw" sounded too slow). The product was renamed BlackBerry. The rest, as they say, is history.

*What's in a name?
It's hard to imagine the
BlackBerry as a Strawberry.*

cupcakes, the package should directly and prominently state "Makes Heart-Shaped Cupcakes." If consumers can't understand what a new product is, they will not buy it.

In addition, depending on your industry and various state and federal regulations, there may be certain information that you *must* include on your packaging (such as ingredients, nutritional content, allergen information, size or quantity, expiration date, address of manufacturer, and so on). Avoid costly mistakes by finding out about these requirements before your packaging goes to print: Contact the Department of Commerce (or the equivalent agency for your state) and other inventors within your industry.

Another aspect of packaging you may need to take into account (particularly if your product appeals to consumers and retailers that may be concerned with environmental issues) is the option of "going green." More and more companies are making the effort to use environmentally friendly pack-

aging (minimal trade dress made with recyclable or easily degradable materials). Some states are even considering legislation to require it, so refer to your consumer research and know your local legislation.

Finding Distribution Outlets for Your Product

As an inventor with a hot new product, your goal will be to find as many *distribution channels* as possible. A distribution channel is an industry segment or subset in which consumers are able to interface with, learn more about, and potentially purchase your product. Think mass merchants (Kmart, Wal-Mart, Target), big-box retailers (Home Depot and Bed Bath and Beyond), specialty retailers (Jo-Ann Fabric and Craft stores for craft products, Sally Beauty Supply for cosmetics), local shops, Web-based marketing (whether it's a personal website or a product sales platform such as Amazon.com), TV shopping channels (QVC and Home Shopping Network), catalogs (The Sharper Image, Sky Mall), kiosks at the mall, and just about any other creative venue you can envision. The idea is to constantly get your product into profitable new distribution channels as often as possible, while avoiding channel conflicts (which could undercut sales through existing channels). Many inventors follow a

The Home Depot may be just the place to reach your product's target market.

progression of selling directly to consumers, then moving business to small retail outlets, and ultimately landing distribution from major retailers—so you can expect your distribution channels to change over time.

Selling Directly to Consumers

Direct sales methods are ways of getting directly to the consumer through distribution outlets such as mail order, Internet, and telephone sales. Selling "direct" can include setting up your own stores (online as well as "brick and mortar"), contacting customers, mail campaigns, catalogs, informational websites, and even direct-response television. Although at first glance selling direct may seem more profitable than retail because it cuts out the middleman, it can be a costly and time-consuming way to reach consumers. Because there are only so many hours in the day, the number of consumers you can reach is finite. Nevertheless, selling direct can be a good way to familiarize the marketplace with your product, particularly in the initial stages of your launch. Direct sales can help you develop market traction (see page 151) and is also a great way to assess initial market response.

DIRECT MAILING Direct mail campaigns are a long-standing method of marketing that now includes direct e-mail campaigns. Traditional direct mailing may be costly (paper and stamps), but it can be effective for particular types

FROM THE ENTREPRENEUR
THE WHOLE PACKAGE

There's a reason most retail stores are seas saturated in bright colors and loud labels rather than a pile of brown bags or blank boxes. Packaging may be the most important element of selling your product to consumers—and of creating your brand. Consumers are often forced to make quick decisions at the point of sale and, as such, must "judge a book by its cover." Your packaging has to be eye-catching and communicate well—after all, it has a lot of other packages to compete with.

of products in specialty markets. For example, catalog mailings for gourmet foods and gift baskets do well around the holidays when consumers are doing holiday shopping and need to conveniently check gifts off their lists.

If you decide to do a direct mailing, you'll need to find out about federal postal bulk mailing rates and other requirements. You may need to hire a service that already has a bulk-mail permit. There's no beating the U.S. Postal Service bulk rates. (Visit their website at www.usps.com/businessmail101/getstarted/bulkMail.htm.) If you're buying a mailing list from a vendor or organization, ask the company how they acquired their list (they often won't tell you, but it is worth asking), and how often they update it. Ask what percentage of the addresses you can expect to be out-of-date. And ask if the company will guarantee that you'll receive no more returns than would be expected using that address

Specialty retail chains, like Michaels craft store, can be valuable distribution channels if your product is the right fit.

inboxes. One creative approach that some inventors have found helpful is to call organizations (a trade union, for example) and offer them a donation (a percentage of your profits) if they will e-mail members on your behalf. The big plus there: Such e-mails tend to be blocked less, if at all.

CATERING TO CATALOGS Catalogs are another way to cut down on direct mail and printing costs. The *SkyMall* catalog, for instance, reaches nearly 2,000,000 air travelers a day and is distributed to airline passengers in the pocket in front of their seat. Customers can order products by phone, mail, or online and are prompted to do so because the catalog is visible during the entire flight. Bored passengers who find themselves without their laptops and other electronic devices are conditioned to look for the *SkyMall*. Passengers know the catalog will be there and that there is always something new and interesting in it. Customers can order from their cell phones while the plane is taxiing, or they can take the catalog and order form with them to order by phone or online.

data. In any event, remember that most unsolicited mail is quickly assessed and discarded by consumers, and even a successful mailing is likely to garner only a 1 to 2 percent response rate. Ask yourself how much visual information potential buyers need: Is it necessary that they see a picture of your product, or is text sufficient? If they need a picture, is two-color or full-color printing necessary, or is black and white sufficient? It's important to gather as much of this information as possible up front so you can calculate costs. The more targeted your list is, the higher the response rate will be, but if the standard response rate won't result in a profit, you might want to reconsider using this method.

Though arguably cheaper, e-mail campaigns can be tougher to launch than you might think. Today spam filters block a large percentage of the unsolicited offers that bombard our

As with most forms of marketing, catalogs are a good way to assess initial market response to your product. Some catalogs—likes *SkyMall*—have a loyal following among people with disposable income and a desire for something unique. Other catalogs cater to special industries, hobbyists, or fashion niches. Nearly any type of product can be sold through a catalog, and catalogs can be an inexpensive way to break into a market and build a following.

The advantage to catalog selling is that while retail distribution channels demand a margin of 50 percent or more of the selling price, catalogs work on a smaller margin (20 to 50 percent). Most catalogs require you to pay only part, if any, of the operational costs of the catalog or catalog company, because the costs themselves are generally split between multiple vendors.

Also, unlike major retailers, catalogs don't discriminate against single-product vendors. While Wal-Mart may require that you have several products to make it worth their time and money, catalogs just want you to have an interesting product that enhances the appeal of their offerings and will be delivered on time.

The disadvantage is that unlike major retailers, who buy your product outright, some catalogs require you to hold the inventory and drop-ship orders directly to their customers. And since catalogs are distributed only periodically, you'll also move fewer products than you would at a major retailer with many outlets.

Most catalogs will review your product to determine whether it's suitable to add to their product line. They'll also consider whether you can ship and deliver on time. If they take you on and you fail to do so, they'll drop you as a partner. (This can be a convoluted problem when your product suddenly sells units faster than you expected, or if demand for it comes in spurts.)

The catalog company typically accepts payments on your behalf and remits them to you within thirty days of shipping. The risk with catalogs is that many of them are small businesses themselves, sometimes operating on a shoestring, and it is not unheard-of for them to go out of business. Do your research to find out how long a catalog company has been in business, either by asking them directly or by asking for back issues of the catalog, and checking with companies that already advertise with them. Then structure the contract (assuming this is part of the agreement and is applicable) so you don't pay for printing until you at least see a copy of the catalog proof (a finished copy of what your ad will actually look like). Don't agree to any arrangement that requires them to be your exclusive marketer unless the company is such a large outlet that it

FROM THE LAWYER
BANKABLE ASSETS

A purchase order, especially from a major retailer, is a bankable asset. You can borrow against it from a bank or a factor (a company that finances projects like this on a short-term basis for a high interest rate). You can also take your purchase order to a manufacturer and convince them to do a joint venture with you. For example, if the manufacturer normally requires payment in net 30 days, you can negotiate for terms of net 90 days which will allow you to pay off the manufacturer with the funds you get from your retail customer. If all goes well, you collect your payment at the end of the process, and bank your profits after everyone else is paid off.

is economically worthwhile to forgo all other marketing opportunities.

The process of getting your product into a catalog usually involves the following steps:

1. Research catalogs that might carry your product. Locate catalogs that carry products that appeal to your target market, and are priced near to your price point.

2. Contact the catalog companies to submit. Once you find catalogs that you think are good prospects, contact them to find out the procedure for submitting a sample of your product (most catalog companies make decisions twice a year or more). A good strategy is to mail them a couple of months before the final submission date, and again just after the date has passed (even if you were rejected the first time). Resending after the deadline may give you the opportunity to fill a hole in the catalog if another product unexpectedly drops out.

3. Check out the catalog company. Before you submit your product, compare the catalog's circulation numbers and printing costs with those of others in the industry. Do a background check by contacting other noncompeting companies that have previously advertised in the catalog, and ask them how satisfied they were with the results of their advertising. In particular, request back issues of the catalog to identify and

contact companies that have dropped out—find out why. Was the catalog company too slow to pay? Did they pay at all?

DOING DIRECT RESPONSE TELEVISION

Direct response television (DRTV) allows you to pitch products through demonstration. It's a powerful form of marketing, since you have the chance to go into millions of homes, talk to consumers, and convince them to buy your product. DRTV requires an average investment of $50,000 to $200,000 (depending on format) to start out, precisely because it reaches so many viewers, but your returns can be substantial and immediate. There are many DRTV companies that will partner with you if they believe your product is a good fit for this form of selling, and they will take a risk on your behalf by assuming some production expenses.

Companies that purchase airtime at a discount from broadcast stations and cable networks manage many DRTV outlets. Unlike general TV advertising, direct response TV ads skillfully ask the consumer to grab their credit card and make the call, which they direct to a well-organized call center. You know almost immediately if your product is a success, based on the responses. DRTV ads are professionally produced, taped, and scripted. It's important to confirm that the people taking orders are well trained and have the inventory on hand to fill your orders.

SkyMall catalog regularly reaches a quantifiable and loyal market.

The advantages of DRTV are that you get the opportunity to demonstrate and explain the value of your product through an infomercial in long format (28 minutes) or short format (60 to 120 seconds). If your product requires instruction, DRTV is a good manner of distribution. For a product to be a good fit, it should ideally support a retail cost that is five to six times the cost of producing it. This large margin is necessary to cover the costs associated with producing the spot, purchasing the media time, and the operational costs of fulfillment (customer service, warehousing, credit card processing, and distribution).

Also, your product needs to have a price point that generally fits in the $9.99 to $49.99 range. Fitness products tend to fetch a much higher price, whereas cosmetics and household gadgets fall within the lower range. In general, though, products priced at $29.99 sell best. (These are products that probably cost around $5.00 to manufacture.) Higher-priced items also sell well using DRTV, but they typically come with special payment programs such as "three easy payments of $59.99."

One particular advantage of the DRTV marketing model is the ability to test various prices, pricing strategies, special offers, and the resulting increases or decreases in orders. For example, a DRTV company might try selling a $99.00 product for a single payment or three payments of $33.00. They might also try offering to let the consumer try the product at home for free and charge the customer's credit card only if they don't return the product in thirty days. Think of how many people are too lazy to return a product they don't want. The marketer realizes this, and as a result, free trials can be a very effective sales technique in DRTV campaigns.

GETTING ON QVC AND OTHER HOME SHOPPING CHANNELS Home shopping channels such as QVC and HSN are a huge boost to inventors who have great products to demonstrate but can't yet afford to take them to "mainstream" television. That's because airtime on QVC, which is the largest shopping network on television, is free.

As with a DRTV campaign, you have the opportunity to have a virtual presence in people's homes and solicit their orders. However, a spot on a home shopping channel doesn't come without its own obligations, costs, and risks. Prior to going on air, you must be able to deliver a substantial amount of inventory to their warehouse for quick order fulfillment, and you must be willing to take that inventory back if the product doesn't sell. (Even if the shopping channel sells your product on air multiple times in order to try to sell through the inventory, you could still see the rest coming back to you.)

QVC has an audience of loyal, affluent, mostly female, repeat customers whose average age is fifty-four. If you have an invention that appeals to this demographic, QVC may consider you as a partner. QVC is always looking for good products to fill 24 hours a day of air-time, 7 days a week, 364 days a year (they are closed on Christmas). Inventors that present a good product and a good fit are usually well rewarded. QVC evaluates products much the way other retailers do: They look for quality products that deliver value to their shoppers.

Of course, not every product is right for QVC. Before you approach them, here are a few questions you should answer first.

> **The Web can be a great place to sell your product, but you won't sell anything if customers don't know you exist. While Web-based sales do increase relative to retail sales every year, most of this activity occurs on websites that already have a strong retail or branded presence.**

■ **Is your product appropriately priced for QVC?** QVC prefers products in the $15.00-and-up range. If your product is offered at a lower price point, it won't be a good fit, regardless of other criteria.

■ **Does your product appeal to the QVC target viewers?** Eighty percent of QVC viewers are women with an average age of fifty-four. And, gender stereotyped as it may sound, jewelry,

cosmetics, and products that provide novel solutions to household problems are a good fit for the QVC demographic; items such as baby products and toys are less so. QVC does not accept items in the following product categories: feminine/personal hygiene, firearms, fuel additives, gambling-related products, genuine furs, sexual aids, and tobacco-related products.

■ **Is your product sufficiently far along to submit to a QVC buyer?** Buyers must have a product that is developed enough for them to evaluate its potential. QVC generally requires you to have samples of a completed product, but they will take a look at a working prototype provided you demonstrate your ability to produce an attractive, high-quality final product prior to issuing a purchase order.

■ **Can you afford to fulfill a purchase order without a guarantee that the product will be sold?** QVC requires that you be able to provide them with sufficient inventory of your product at wholesale cost (about half of what the item will sell for on the show). The product is shipped in advance of your airdate to one of their warehouses. Even though you've received a purchase order from QVC, the product is still subject to meeting minimum sales performance on air. If a product sells well within its allotted time slot, you can expect repeat appearances. However, failing to meet these minimum sales requirements could lead to the unsold product being returned to you (at their expense).

If you believe QVC is a good fit for your product, you can present it at one of their road shows, by contacting the online product submission department via their website at www.qvcproductsearch .com, or attending an *Everyday Edisons* casting call (representatives from QVC travel to each of the casting calls). QVC's online submission process is accessible and straightforward. After you complete the submission form, you'll immediately receive an e-mail confirming your submission and providing you with a product submission number. The confirming e-mail will also request that you attach and return a digital image of your product. QVC will then contact you in four to six weeks to let you know their level of interest. At that point, if they're interested in your product, they will request samples and set up a meeting. They prefer that you do not send unsolicited product samples; they will give you mailing instructions if they request samples.

Although the process of submitting to QVC is relatively quick and painless, it may seem a bit impersonal. Remember, though, that they receive hundreds of thousands of product solicitations every year. Tamara Monosoff, author of *The Mom Inventors Handbook,* recalls that her first QVC selling experience was not without its misunderstandings. Shortly after she submitted her first product for review, QVC e-mailed her to say they were interested and invited her to pre-sent her product, the TP Saver. She was elated. She arrived to her presentation feeling quite special, only to find 2,800 other "special" inventors waiting in line to present their products!

In her book, Monosoff assures readers that "QVC is a well-oiled machine, and while I thought the cattle call experience would be frustrating, the staff members treated each of us as if we were the only one in the room." Her first product didn't meet their price criteria. At the time, the preferred minimum unit price for QVC was in the $15 range and her product had a retail price of just $5.99. But she persisted and has since graced the QVC "stage" several times.

SELLING ON THE INTERNET: A WORD ABOUT WEB SALES The Web can be a great place to sell your product, but you won't sell anything if customers don't know you exist. While Web-based sales do increase relative to retail sales every year, most of this activity occurs on websites that already have a strong retail or branded presence (think Amazon.com or Buy.com). We recommend that you pitch your product to these distribution outlets much as you would if you were going after a purchase order from a brick-and-mortar distribution outlet and leave the Webmaster role to them.

It took the Amazon.com online marketplace nearly ten years to become profitable, when its brand as a go-to site for books and products finally took hold. Today Amazon generates billions of dollars in sales and is larger than most brick-and-mortar stores. Launching your own website and trying to get it to surface in the results when consumers search for your product can be a catch-22. If it's a frequently used search term (such as "computer accessory"), lots of competition may come popping up

---

Your website should present your product well: Convey an overall image of quality and brand value, just as you would on a store shelf.

---

ahead of you, and your expensive website may land far down on the search pages. On the other hand, if your product is truly novel and different from the competition's, it will land high on the lists *if* customers know the keywords to use to search for it.

If you are able to overcome all the barriers of entry to getting your page to come up on consumer search results, Web sales do have a certain advantage, primarily because they generate a high profit margin. Websites sell directly to consumers and retailers, cutting out distributors.

One way to increase your online presence is to purchase "ad words" that cause your website to come up in the "paid search-results" section when consumers enter them in search engines. For example, if you sell a scrapbooking product, you might purchase the phrase

"scrapbooking." Your website link will surface in the paid search-results section anytime someone searches that word. You pay for this form of advertising each time users visit your website and the amount of your payment is determined by how many others "bid" for those terms. This can be an expensive form of advertising, and the results, depending on the category, will vary based on what type of product you have or what problem you solve. For example, an online store that sells used books may have a lot of competition, whereas a site that sells a cure for an exotic ailment may do quite well in relationship to its niche market.

Rather than sinking a lot of money into Web sales, we recommend that you consider your website part of your invention's overall branding program. It is a place where people can get more information about your product and where they can order it when they see it advertised. Your website should present your product well: Convey an overall image of quality and brand value, just as you would on a store shelf.

Taking Your Product to Trade Shows

Just about every industry has a trade show, and if your product is industry specific, you'll want to attend at least one show. For example, if you have a hockey-related invention, you'll want to go to the Let's Play Hockey International Expo held every January in Las Vegas. If you have a hardware or home-improvement device, you'll want to attend the National Hardware Show. Trade shows can be expensive, and you have to decide which ones are worth attending, and then whether to attend them as an exhibitor or to walk the show as a buyer. One inventor, Daryn Reasby, never attends trade shows as an exhibitor. His principle is that, as an exhibitor, "you're locked into your booth space"; he prefers to be more mobile and talk to people in the aisles. Reasby carries a sample of his product with him or garners attention by walking down the aisles with his miniature remote-controlled jeep cruising alongside him. He makes more contacts than he can count. The advantage of just attending the show is that the cost is significantly reduced. Booth fees for exhibitors at a show can cost thousands of dollars, if not tens of thousands.

If you do decide to walk a show, make sure this practice is allowed and be respectful of the vendors who have paid to exhibit. Soliciting manufacturers during a busy show is frowned upon. When a show is slow, especially early in the morning and late in the evening or on the last day, many manufacturers will welcome dialogue with potential vendors or inventors.

For some inventors, trade shows are exhausting, since an exhibitor must stay within the confines of a booth. Others who have a novel and eye-catching product, or have found dramatic ways to showcase their product, enjoy the reactions of the people at the show. If

FROM THE ENTREPRENEUR
DON'T STEAL THE SPOTLIGHT

At a trade show, ease yourself into the crowd and introduce yourself to as many people as possible—your goal is to make friends. Work your invention into a conversation; and make sure you're not stealing the show from the people who paid and set up appointments to be there. If you want to call a lot of attention to yourself, do so near the end of the day, when people have already gotten in a decent amount of booth time.

your product can be demonstrated well within the confines of a booth to passing crowds (such as battery-operated toys that do something entertaining, or home décor items), exhibiting at trade shows may be a worthwhile investment. However, if your booth consists of a table with a bunch of brochures, you may be better off walking around and making contacts as an attendee. Like any other form of marketing, this should be evaluated for the return on investment: Can you, as the exhibitor, generate enough revenue from the event to justify the cost of attending?

Personally Pitching Your Product to Local Outlets and Beyond

Many successful products start locally with relatively small retailers, long before they land that large purchase order. Many local retailers like to buy local

Trade shows, such as INPEX (The Invention and New Product Exposition), shown here, can be exhausting but rewarding for some inventors.

products and are looking for unique items to stock their shelves and differentiate themselves from the big-box retailers. So scheduling appointments with independent, locally owned retailers is a good way to build objective evidence of the sales potential of your product. It's to their advantage to buy directly from you, rather than a sales representative, because it cuts their costs. In general, if you have a good product that can be profitable to them, they'll be happy to meet with you.

As you perfect your sales technique and earn product credibility from selling to local retailers, you'll be ready to move on to increasingly larger distribution outlets. Regardless of the size of the company you're pitching, the best approach is to find the decision maker within the company. In a small company, this may be the owner or CEO. In a larger one, it will be the general mer-

chandise manager or category buyer. The largest retailers employ separate buyers to purchase products in very defined product categories (electronics, toys, books, and so on).

Plan your *cold calls* (unsoliciated sales calls) carefully. Yes, it *is* a numbers game, but you'll be far more successful if you make well-prepared, targeted calls rather than random, scattershot ones. Find the appropriate person to contact, and see if you can find out some small point you have in common. It could be a hobby, an alma mater, or a common cause you both support. Conduct an online search or visit social networking sites like LinkedIn.com and Facebook.com. You just might find your contact there. After introducing yourself, make that connection and then succinctly describe your product and what you do. Speak slowly and in a casual, friendly tone. Be concise, and

Pitching to the Toy Industry

Individual inventors dominate the toy industry, even though only a small percentage of ideas are launched. In this fickle industry, with the rapid-fire dynamics of electronics, computers, and pop culture coupled with adult desire for nostalgia, toy companies simply cannot afford to ignore individual inventors in keeping up with consumer tastes. Toys and games are not just for kids. There is a huge fun-loving toy and game market that spans generations.

Software developers Whit Alexander and Richard Tait caught a lucky break when their "self-published" (or bootstrapped; see page 171) game caught the attention of Howard Schultz, the CEO of Starbucks. The brainy board game Cranium was a perfect fit for the Starbucks target market: younger, hard-working, coffee-guzzling professionals (to generalize); and the launch within Starbucks stores gave the creators of the game an enormous distribution channel to get the product in front of consumers. Based on that initial launch, the game is now sold by virtually every major retailer that sells toys and games.

If you are playing with the idea of launching a toy or game, keep these tips in mind.

1. Put money into your prototype and packaging. Toy buyers are especially conscious of the visual appeal of the finished product.

2. Invest in a booth at a toy or hobby show, or simply attend. Most toy buyers attend several shows during the year to look for new ideas.

3. Keep track of trends; the toy market is fast-moving.

4. Keep the basic concept simple. Getting a buyer to focus on overly complicated items that are difficult to understand requires more effort.

5. Protect your idea with a patent. The problem with simple (albeit ingenious) concepts is that they are easy to emulate. Most patent attorneys can suggest economical strategies such as the simplified application known as a provisional patent application (see page 113) to protect you in the first year of your product launch.

Cranium, the family board game, was first launched in Starbucks stores.

avoid the aggressive pitch. Do not try to spit out the entire sales pitch in thirty seconds, but refrain from rambling on. Focus on selling the uniqueness of your product, the value proposition, and why you feel the product is a good fit for them. The goal is to set up a meeting where you can spend thirty minutes selling the product in person.

The ease with which you maneuver the conversation to a comfortable point will build credibility for you. Research and preparation will help you warm up the cold call by finding a small point of rapport; it will be well worth your time. For example, instead of saying "I'm calling about a sporting goods product," try something along the lines of "Like you, I am an avid athleticist, and I've developed a product that makes practicing a sport even more enjoyable . . . ," submitting specifics (cyclist/cycling, soccer player/soccer) where applicable.

Once your product is ready, and you've developed a solid base of smaller accounts, you're ready to approach the big-box stores.

Also, be sure to do your homework and focus on the company's needs rather than the abstract benefits of your product. What type of products does the company sell? To whom does it cater? Why will your product be a good fit, and why will the company's customers be glad if the company decides to carry it? Why will they find your pricing appropriate? Rather than saying "My product is a fabulous value," try "I see your company sells ABC, and we have a product that appeals to consumers in that price range."

Offer to send samples, photos, or brochures. Better yet, drop them off in person. If you are fortunate enough to get an order immediately (sometimes buyers are so busy they will do this), thank them sincerely and end the call.

Duplicating Yourself: Hiring Sales Representatives and Agents

An independent sales representative (independent rep) is someone who can represent you in promoting your product to retailers, and who helps you service those retail accounts. Many inventors act as their own agent for accounts near home and enlist independent reps to help them expand into other territories. Reps often carry other products in addition to yours and have their own lists of retailers to whom they sell.

Independent reps usually work on commission (from 3 to 10 percent of the net selling price); consequently they take on only products they think they can sell. They may represent your product to small local merchants, regional chains, or large national retailers. If a rep asks you for a lump sum up front to represent your product, it's a red flag. Set minimum sales expectations and avoid granting exclusive rights to a rep unless those expectations are realized. Regardless of the services your sales agent is providing, you should always have a written agreement addressing:

- **Commissions:** These are normally 3 to 10 percent of the net price, based on total order minus freight, discounts, returns, and allowances.

- **Cost reimbursement:** Generally there is none; the exception would be certain travel or other expenses that you require the rep to incur and agree to reimburse up front.

■ **Identification of entities:** Identify the types of entities to which they may present your product, and geographic territory, such as suntan lotion in a resort area.

■ **Terms of agreements:** How long will you engage this sales rep? What duties do you expect them to perform, and how often will they be compensated?

■ **Purchase order terms:** Are they authorized to negotiate on your behalf, such as discounts and shipping dates? Will you set pricing that they must use?

■ **Minimum expectations:** Set sales performance standards to keep the terms of the agreement in effect or to maintain their exclusive territory (if granted).

Some retailers may have "vendor consolidation" policies which require them to use as few single product vendors as possible to minimize paperwork. In addition, many companies (particularly buyers in larger retail stores) are more confident when they're working with an experienced representative who has a track record in bringing them multiple successful products than with individual inventors pitching single products. Since the buyers themselves are taking a risk, they may feel that if the representative has taken on the product, the rep has already verified its quality, your ability to deliver, and other matters of that kind. On the other hand, many inventors choose to perform the "representative" function on their own behalf. In the end,

the decision comes down to returns and results, and the best way to effect both.

Getting into the "Big-Box" Stores

There is no disputing the lure of large retailers. Getting your products onto the shelves of a mass merchant (Wal-Mart, Target, Kmart) or a big-box retailer (Home Depot, Staples, PetSmart) certainly isn't right for every business, and it is not the only route to retail success. But if your products and company are a good fit for major retail outlets, supplying to them can put you on the commercialization fast track. Landing distribution with a major retailer with thousands of stores, puts you instantly on the map.

If you're a small inventor and/or small business, a single large purchase order with a major retailer may catapult your product to success. Securing your product a place on the shelves of a major retailer means you've reached national distribution, but along with this comes a completely new set of issues.

GETTING IN THE DOOR But how do you get there? Once your product is ready, and you've developed a solid base of smaller accounts, you're ready to approach the big-box stores. In order to convince them to take you on as a new vendor, you must demonstrate, as with any buyer, that you have a good product that will generate sales and that you have the ability to satisfy their purchasing requirements.

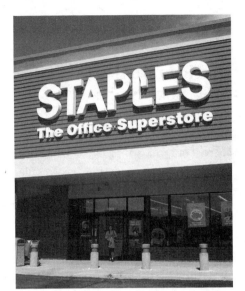

Placing your product in mass merchant and big-box retailers can put you on the map.

a unique identifying number for your product, similar to a social security number. See page 74 to learn how to obtain one for your product.

■ **Product liability insurance:** You must have insurance in place before major retailers will assume the liability of carrying your product in the event that a consumer is injured because of it.

■ **Names of other accounts to whom you've supplied your product, to show that you can ship on time:** This establishes your credibility. Major retailers want to know that you won't go under if they decide to change their product line. They want to see that you have a solid track record and history of supplying your product, and usually that no more than one third of your total business will be coming from them. There are always exceptions, but they don't want to be your first account, and they don't want to be your major account. Wal-Mart asks for references and names of other major vendors (specialty chains or big-box stores) to which you've supplied your product in the past (if you have them). If you don't have them, be honest, and address directly any concerns about your ability to deliver a product on time (for example, by citing your track record with small retailers). You should include these references in your proposal to all other major retailers as well.

To get a meeting with Wal-Mart or its affiliate, Sam's Club, start by filling out their online vendor application, which outlines the process for preparing to present to them. Visit www.walmartstores.com for more information (click on "Suppliers" to reach the correct page).

Other stores, including Home Depot, Target, and Costco, don't have standard forms; rather, they invite you to e-mail them a proposal. Occasionally these retailers will have open vendor days when any new vendor can come and meet with a buyer. Use the Wal-Mart online application form as a template for submitting your product to land a meeting with any of these buyers.

Assuming your product is priced right for these major retailers and fills a niche, here are a few things you will need to gather up before submitting a proposal.

■ **A Universal Product Code (UPC):** As discussed in Chapter 3, a bar code is

ACING YOUR MEETING WITH A BUYER

Every buyer for a major retail store sees a new vendor as a risk, and there is an inherent bias against new vendors. Will they deliver the goods and deliver them on time? Buyers' careers depend on profitably stocking store shelves with products that people buy. Quick turnover means profits. The simple truth is that for a retailer to purchase your product, they must *not* purchase someone else's. It's a trade-off, and sometimes it's just easier for buyers to keep buying what they have bought in the past. The reason they will consider your product is that they believe your product could sell better (more units) than what they are already purchasing.

If you successfully complete the application submission process and land an interview with a major retail buyer, here is a checklist of things to remember.

■ **Buyers see a lot of products.** Don't take it personally if they seem bored or are rushed; proceed with confidence. Never assume you have more than eight minutes to present; prioritize your points accordingly. Focus on what makes your

Breaking into Big-Box Stores

Big-box stores are large chain retailers whose buyers see a lot of products, most of which are polished, packaged, and ready to go. However, stocking the store shelves is not as cushy a job as it may seem. Buyers are held accountable when vendors miss delivery dates or consumers complain about quality. It's no wonder that they are skittish about single-product vendors and new inventors with blank spaces for references.

If you fall into these categories, here are some ways to build buyer confidence.

1. Build your résumé. Start your marketing strategy by building a solid track record with local retailers. Use them as references and be sure to earn rave reviews.

2. Generate buzz. Find a way to get consumers to take notice of your product. You can generate PR in newspaper articles, TV-show appearances, or even online chat rooms or blogs. If your budget allows, advertising is also seen by retailers as being supportive of sales.

3. Expand your product line or team up with other inventors. It is a lot of work for a big-box retailer to process paperwork for a new vendor. Consider launching a number of products, teaming up with other inventors, or hiring a sales agent.

4. Be prepared for the pitch. Expect lots of questions about your manufacturing and distribution capability and be prepared to address them.

5. Demonstrate flexibility and a willingness to do what it takes. Whether it's a request to modify your packaging or to bend on your pricing, you have to show you're willing to work with the retailer whenever possible, as long as it makes sense for you and the success of your product.

product unique, the reason consumers prefer your product, and the reason the retailer should purchase it.

■ **Buyers are looking for products that fit within their current line.** Succinctly and confidently explain why and how yours fits that niche. Assume the buyer doesn't have or doesn't remember any of the materials you previously submitted.

■ **Come prepared to offer more than one product or to enter into a relationship with a sales agent who does.** Setting up a vendor takes considerable paperwork, and major retailers generally have stringent rules about dealing with single-product vendors. Even if your product is fabulous, having only one product to offer could be a deal

killer. Suppliers who can offer a complete line of products (or sales agents presenting multiple products) generally have an edge.

■ **Show that you can ship on time.** Although it takes an average of six months to a year for a supplier to get a first purchase order, you're expected to move fast when that first order comes in. Often the retailers want turnaround in a matter of days or weeks.

■ **Price it right.** Most retailers look for margins between 40 and 65 percent. Find out what margin the particular retailer looks for in your category. Make sure you can supply them the product at the price they need in order to be profitable, while allowing yourself to be profitable, too!

CHAPTER 6

Doing the Math

Licensing vs. Manufacturing

How much is a good, patentable idea worth? The answer depends on an inventor's ability to get his or her idea to market himself or to find someone who can do it for or with him.

Back in 1921 a fourteen-year-old Mormon farm boy named Philo Farnsworth had an idea while working on the family farm. Young Farnsworth was a brilliant math and physics student, and one day, caught in the monotony of his chores, he was suddenly struck with the realization that an electron beam could scan a picture in horizontal lines almost instantly.

Philo Farnsworth's combination television and radio receiver, 1935.

In 1927 when he was just twenty-one, he filed a patent application on the television set.

Farnsworth's early television was a camera that focused an image through a lens onto light-sensitive cells in a glass tube. The electrical image formed by the cells in the tube would be scanned by an electron beam line-by-line and transmitted to a fluorescent screen. It worked amazingly well, but Farnsworth didn't have the funds to develop the invention or bring it to market. Moreover, he didn't possess the temperament to woo investors or manage a team. His invention languished for a decade, as Farnsworth became embroiled in vari-ous legal and management squabbles, neglected his family for his ideas, and grew increasingly depressed.

Finally, in 1934, an acquaintance put him in touch with a British company called British Gaumont. British Gaumont entered into a license agreement with Farnsworth to make a basic television system using the technology covered in Farnsworth's patents. Five years later Radio Corporation of America (RCA) followed suit and also entered into a license agreement with Farnsworth for his patent rights, in order to begin competing with Gaumont. The relationship was famed for its acrimony: It was no secret that RCA president David Sarnoff and Farnsworth disliked each other; Sarnoff had once mounted a legal challenge to Farnsworth's patents and lost. Both companies, however, were required to deal with Farnsworth because, as the patent holder, he had the exclusive right to prevent others from making, using, or selling his invention. Both RCA and British Gaumont began developing commercial versions of the television set, building on Farnsworth's technology while competing with each other. As a result of their licensing agreements with Farnsworth, regular television broadcasts became available to the American and British public after World War II. Farnsworth's story is an important example of what licensing can accomplish for inventors, companies, and consumer markets.

The Market Dynamics of Licensing

Getting a finished product into the hands of paying customers requires more than a great concept. It requires product design, engineering, manufacturing, packaging, advertising, shipping, compliance with laws and regulations, and most important, enough money to do all these things. Many inventors find themselves with a well-researched product and an untapped market, wondering if they should use their personal savings to get that good idea to the public or minimize their risk through licensing in exchange for more limited profit potential.

Bringing a New Product to Market: A Calculated Risk

Regardless of the size of a company, introducing a new product is always a risk. You can't guarantee that consumers will conform with market research. Market research, after all, is merely a prediction. For example, take Coca-Cola, one of the most recognized brands in the world and a company that invests hundreds of millions of dollars in launching new products. In 1985 the Coca-Cola Company decided it needed to generate more sales for its flagship Coke brand. After spending millions of dollars on development and holding countless focus groups, the Coca-Cola Company was sure it had a winner when it introduced New Coke. But rather than boost the company's sales, the introduction of New Coke resulted in public protests and demonstrations with bottles being emptied into the streets of southern cities. Coca-Cola became the butt of late-night talk-show monologues, scathing editorial columns, and even a consumer lawsuit. The "old" Coca-Cola formula was quietly reintroduced. The Coca-Cola Company had made a calculated risk and had grossly miscalculated.

Assume for a moment that your product *will* be a huge success, and has the potential to take market share away from competing products. Assume, too, that the cost to bring the product to market is fairly reasonable and that you have the ability to raise the necessary capital and the skill set to run a business. You certainly stand to make more money by manufacturing it yourself than by licensing the idea to someone else. This is the ideal situation. In reality, there are no certainties that a product will prevail, and no guarantees that consumer response will be what you predict. So whoever assumes the costs of developing and manufacturing a product is assuming the greater market risk, for which they are entitled to greater compensation.

If you're an individual inventor with one product (or maybe even a few) in your portfolio, doing the manufacturing yourself places you at the mercy of an unpredictable marketplace. Large companies, on the other hand, are positioned to weather shifts in consumer

The Coca-Cola Company took a risk when it launched New Coke in 1985.

CASE STUDY
A Better Way to Take Out the Trash

Franklin Ramsey of Charlotte, North Carolina, considers himself both a family man and an "idea man." Years of caring for the family home have put him in a position to brainstorm easier ways to carry out the menial, time-consuming tasks of life. Ramsey developed the idea for his invention, a trash-can technology now known as Pressix, while helping his wife, Ann, with her fledgling commercial janitorial business.

When Ramsey tried to empty wastebaskets, he could never seem to secure new trash bags properly. His knotted bags were either too loose or too tight, or they tore as he was trying to make the knot. Frustrated, Ramsey proceeded to his garage workshop and devised a trash can that holds bags securely in place without the slipping or ripping. Shortly thereafter, he heard about the *Everyday Edisons* casting call in his area and decided to get some expert advice.

Ramsey so wowed the judges with his straightforward yet savvy trash-can invention that his idea was chosen as one of the fourteen for the show. Pressix was eventually licensed to two leading manufacturers of decorative garbage cans, Umbra and Home Zone, and was unveiled at the Chicago International Home and Housewares Show in the spring of 2008. The reaction to the product was overwhelmingly positive; Home Zone secured orders for more than one million trash cans that incorporate the Pressix technology.

Franklin Ramsey created an initial prototype for his invention by using a box cutter to modify a standard trash bin.

trends and tastes; they can adjust, in part because they are able to spread their risks and failures over many different products. In addition, they often use the information they get from failed ideas to develop new ones (individual inventors do this too; large companies simply have a wider reach of resources).

Larger companies may also have the ability to deflect issues that individual inventors can't anticipate early on. Companies that have already brought comparable products to market may have specialized industrial-design or engineering expertise, and they may have maneuvered regulations, safety standards, and difficulties that you might not see in manufacturing a product until you are well into the process. Most important, though, they have relationships with retailers, distribution channels in place, and track records that give them "street credibility."

Suppose you are bringing an automotive product to market. Do you have the ability to spot and reasonably identify the barriers you face? Your prototype may work perfectly but still not be as far along as you think. Consider the case of Franklin Ramsey, one of the inventors featured on *Everyday Edisons*. Ramsey's wife runs a janitorial business, and one day she asked him to assist her. His job was to empty the trash cans and replace the full bags with new liners. If you have ever put a plastic liner in a garbage can, you know it's a challenge to get the liner to fit properly. Typically, you need to tie a small knot in the bag to get it to stay in place. After several failed attempts to line the containers, Ramsey decided there must be a better way. When he got home that evening, he took an old plastic garbage can and began making cuts in the side to create a bag-locking feature.

Once he had a prototype, he didn't know where to turn. As an individual inventor, Ramsey was hard-pressed to assemble a team, fund the research, or know how to pitch a product to retailers. Large companies can offer this expertise if they license your product, and can often deal with obstacles long before they become a problem. Even the most ingenious idea can use a little help sometimes.

The Licensing Trade-off

Profiting from your invention is about risk, reward, and understanding the balance between them. Licensing is an example of that trade-off. It is a way to profit from an invention while minimizing the personal risk and commitment. As with any investment in the idea-to-sale process, since there is less risk, there is also less profit potential. The upside, or reward, is a royalty paid on sales.

Your primary initial investment in the licensing relationship is to protect your idea by filing a patent application on your product, and to develop, at the very least, a proof-of-concept prototype that demonstrates the usefulness and function of your idea. You, as the inventor, act as the *licensor*, and you try to find a "licensee" who will develop your idea from there, consulting with you along the way as needed.

In exchange for the right to produce your patented or patent pending product, a licensee will pay you a share of the *revenue* (money collected from sale of the products). This payment is called a *royalty*. On average, royalties for consumer products range from 2 to 5 percent of the total product revenue collected by the licensee. There are, of

FROM THE LAWYER
LICENSE TO CREATE

Licensing can be a great way for an individual inventor to profit from an invention without risking a lot of funds up front to develop manufacturing and distribution capability. However, it is almost always less profitable than manufacturing a product. The real rewards generally go to the party who has the financial wherewithal and is willing to assume the risk of actually bringing the product to market, as opposed to merely conceptualizing it. When you enter into a license agreement, you generally agree to take a small share of the profits from a product you let someone else produce using your patent rights. Manufacturing often yields profit margins in the 50 percent range. A typical royalty is 2 to 5 percent of the sales price per unit.

Let's Get Serial

Turning a great idea into a product that generates an economic reward is a wonderful feeling, but wouldn't it be even better if we could actually invent for a living? Here are a few *serial* inventors who have done just that.

Dean Kamen A modern-day Edison with more than 440 patents, Kamen is responsible for such inventions as a portable dialysis machine, iBOT (a stair-climbing mobility device), and the infamous Segway scooter.

Kane Kramer In 1979 at just twenty-three years old, he came up with the concept for a portable digital audio player. In addition, he developed technologies for downloading music and created a solid state digital recorder and player.

Jerome Lemelson People either love him or hate him, but the fact is Lemelson amassed a billion dollar fortune by licensing more than 600 patents.

Judah Klausner A serial inventor, Klausner developed the technology behind PDAs and electronic organizers, as well as some voice-mail technologies.

John Osher A serial inventor and entrepreneur, Osher has made a nice living for himself by developing products and either selling or licensing these inventions. His biggest success was the Dr. John's SpinBrush (later renamed the Crest SpinBrush), which he sold to Procter & Gamble for $475 million.

course, exceptions. Royalties can vary based on the market size, expected margins, development cost, and strength of the intellectual property. Assume, for example, that you invent a new widget that you decide to license to a manufacturer. The product sells in stores for $10.00 and was sold to the retailer by the manufacturer for $5.00. Your royalty, assuming you negotiated 5 percent, would be 5 percent of $5.00 per unit, or 25 cents per unit. That's not a huge amount of money, but balance the amount of risk you are assuming against the amount of time and resources that the manufacturer has invested in selling your product.

On the other hand, if you choose to manufacture the product yourself, you may make ten times the amount of income. If you are able to manufacture the widget for $2.50 and sell it to the same retailer for $5.00, you would earn a $2.50 gross profit on every unit sold. There is a greater reward here for you, but also a significantly higher risk. To get that reward, you have to make the investment in manufacturing, inventory, and all the costs associated with selling and supporting your product. Although the profit potential from manufacturing your product on your own rather than licensing it is definitely greater, licensing is a unique type of opportunity where you can make money from a product with a very limited initial investment. Your primary outlay will be for obtaining basic patent protection and creating a prototype. These costs are only a fraction of the investment you would make in acquiring equipment, staff, packaging, expertise, shipping, and quitting your day job to run a business.

What you will actually be licensing to the company is not your naked idea but rather your patent rights to exclude others from producing your product. Companies do not license ideas or products, they license intellectual property rights. In order to license your concept you must have a patent application filed (or better yet, an issued patent). Having some form of intellectual property (patent or patent application) is required by virtually all companies that license products, except in very rare instances when an inventor has a valuable, well-known trademark or protectable trade secret. (Since trade secrets must be maintained in complete secrecy, they are difficult to license and very rare.) The cost of filing a provisional patent application to get you started, as discussed in Chapter 4, is usually no more than a couple thousand dollars.

Licensing frees up a lot of inventors. Handing off the production responsibilities can allow you to go on to do what you do best: invent. Many "serial inventors" actually make their living as inventors. They are constantly filing patents on their products and licensing them to companies eager for new ideas.

Deciding to License, Manufacture, or Do Both

The licensing-versus-manufacturing decision is a very individual one for every inventor. It may be hard to hand off a product you are certain will be a great success to someone else who will get the lion's share of the profit. But what if your product, although ultimately successful, takes consumers longer than expected to try and then accept? Can you calmly and confidently get through this period, or will it take its toll on you? It took years for consumers to accept Band-Aids, Teflon, Velcro, and Tampax. These products were introduced to the market by large companies that had the wherewithal to wait out the consumer learning curve. The products have been immensely profitable over the long term, but would have certainly failed if the resources did not exist to support them until they caught on with consumers.

Inventions are as different as fingerprints, and so are the dynamics that determine whether an inventor is better off licensing or manufacturing. Explore both options early in the process, without ruling out either course of action too soon. Many inventors end up doing some combination of the two by switching strategies at appropriate points. For example, an inventor may manufacture a limited amount of product in the early stages and generate some sales to demonstrate its market appeal. This is known as developing *market traction* (or

US00D327726S

United States Patent [19]

Lowrance et al.

[11] **Patent Number:** **Des. 327,726**

[45] **Date of Patent:** ✱✱ **Jul. 7, 1992**

[54] **FISHING ROD BUOYANT**

[75] Inventors: **James M. Lowrance**, Elk City, Okla.;
Johnny W. Hall, Vernon, Tex.

[73] Assignee: **Lowmac, Inc.**, Elk City, Okla.

[**] Term: **14 Years**

[21] Appl. No.: **489,653**

[22] Filed: **Mar. 7, 1990**

[52] **U.S. Cl.** **D22/139; D22/146**

[58] **Field of Search** D22/139, 134, 117, 145,
D22/146; 43/44.91, 18.1, 44.87, 431, 25, 44.9,
17, 17.2, 17.6, 43.14, 43.15; 441/8

[56] **References Cited**

U.S. PATENT DOCUMENTS

1,787,862	1/1931	Hornke	43/43.1
1,836,836	12/1931	Chobanoff	43/43.1
2,741,865	4/1956	Devoti	D22/146 X
2,885,817	5/1959	Carter	D22/146 X
3,038,375	6/1962	Gansz	43/43.1 X
3,570,163	3/1971	Conder	43/17.2
4,583.314	4/1986	Kirkland	43/25
4,766,690	8/1988	Troha	43/25

Primary Examiner—A. Hugo Word
Assistant Examiner—Doris V. Coles
Attorney, Agent, or Firm—Glen M. Burdick; Bill D. McCarthy

[57] **CLAIM**

The ornamental design for a fishing rod buoyant, as shown and described.

DESCRIPTION

FIG. 1 is a left side elevation view of a fishing rod buoyant, showing our new design, the broken line showing of environment is for illustrative purposes only and forms no part of claimed design;

FIG. 2 is a rear elevational view thereof;

FIG. 3 is a front elevational view thereof;

FIG. 4 is a right side elevational view thereof;

FIG. 5 is a left side elevational view thereof;

FIG. 6 is a top plan view thereof; and,

FIG. 7 is a bottom plan view thereof.

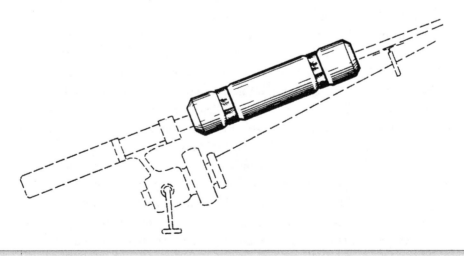

Having enjoyed previous market success, Jim Lowrance was in a stronger position to negotiate a license and royalty rate for his Rod Floater.

evidence that your product will gain popularity with consumers over time).

Consider the story of Jim Lowrance. Jim and his brother-in-law started off manufacturing one of Jim's own products, a flotation attachment for fishing rod-and-reel combinations, and eventually shifted to licensing the product to leverage the distribution potential of a larger company.

Lowrance describes his invention as an "8-inch, cylinder shaped piece of poly-foam material, similar to the kind of material you see used in water noodles and Nerf toys. The cylinder attaches to the fishing rod, just above the rod handles"—it's aptly called the Rod Floater. Lowrance knew little about invention marketing and contacted an invention promotion company. The brothers-in-law sent their invention concept to them in detail, on paper. The company responded promptly, assuring them that the invention had very broad potential and that they had already found a company to license it. After signing several contracts for this corporation to manufacture and market the product and paying thousands of dollars to the promotion company, Lowrance and his brother-in-law learned that the invention had never been licensed as they had been told. They managed to obtain a release from their contract with the promotion company (but did not get a refund).

Sure that their invention still had potential, Lowrance and his brother-in-law pursued marketing it on their own.

They began knocking on doors and were able, through this process, to get several local business people to invest in the Rod Floater. They formed a little corporation called Low-Mac, Inc., and had the invested startup capital to get the Rod Floater patented, packaged, insured, and marketed.

They were then able to get the Rod Floater into a number of regional outlets. In fulfilling these orders, Lowrance learned how to make prototypes and promotional ads, design packaging, and procure the product liability insurance required by many outlets.

Lowrance and company managed, on their own, to sell the Rod Floater to Wal-Mart (regionally), Bass Pro Shops, Cabela's, and Academy stores and to market it on national TV shows. They were even able to secure a national premium-sale deal with a major oil company, who promoted their outboard motor oil using Rod Floaters as a giveaway with their product. In 1996 Lowrance decided that it was in his and his investors' interest to enter into a licensing agreement for the Rod Floater. The agreement enabled him and his partners to leverage the large, national distribution outlets of the licensee and ultimately to increase their overall sales. But without the traction they had attained by initially marketing the product on their own, the negotiating position with the prospective licensee would not have been as strong, credible, or as lucrative. More than a decade later, Lowrance still receives royalty checks. Having market traction enhanced his bargaining position for negotiating a royalty rate.

Weighing the Factors

In the early stages of getting your product to market, you'll be researching the same information whether you are licensing or manufacturing. Regardless of the path you choose, you'll need to know your target market, projected sales, and costs to produce the product.

Ultimately, the decision of whether you go into the business of producing your product yourself or licensing your rights to someone else usually hinges on some variation of the following factors:

■ **Your risk tolerance:** How much of your personal funds can you afford to risk? It takes time and financial resources to bring an invention to market, and if you're not investing at least some of your own capital, others will wonder why they should invest theirs.

■ **Your commitment and skill set for starting a company:** While it takes ingenuity and creativity to come up with an idea for a product, it takes a business (either yours or someone else's) to get that product to market. Do you have the skill set to start a business? Do your talents lean more toward creative engineering than analyzing financial statements and preparing market projections? Investors who help you start a business will assess your management capability and your long-term commitment to the business in the same way as they will evaluate your concept.

■ **Your ability to raise capital and deal effectively with outside investors:** Are you comfortable recruiting friends, family members, and complete strangers to convince them to back your idea financially? Can you come up with reasonable and accurate projections and the data that you need to convince others to invest? Can you offer your investors a higher return than they could get on other investments? Are you comfortable giving those investors equity interest in your company or a say in the market strategy? What will happen if you fail and cannot pay them back?

■ **Your personal goals:** Are you ready to make a major change in your life and disrupt your current career path for what may be a year or more to bring a product to market? Can you work whatever number of hours the success of your company or invention requires? Can you survive without a consistent paycheck or any paycheck at all, and can you live with failure?

If these factors sound abstract, let's put them into context by considering the case of Jeff Cavette, a firefighter who invented a device for him and his fellow firefighters to transport their gear to and from firehouses in a way that does not require them to stuff their wet, smoky coats, helmets, goggles, and boots into a duffle bag after a fire. His invention is a sort of strap that allows firefighters to quickly gather up, organize, and carry their gear in the first critical moments after a fire alarm sounds, reducing the risk of leaving an essential safety item at the firehouse.

Cavette initially took his newly patented device to firehouses to survey firefighters and gather them into focus groups (using the basic market research techniques discussed in Chapter 2). These groups—his target market—gave his product rave reviews. They decided that it was a product that would be offered at a price they were definitely willing to pay (as long as the product was manufactured domestically). Since there are more than 1.1 million firefighters in the United States alone, early indicators were fairly convincing that a market and some profit potential existed for Cavette's fire gear device.

The simplest invention can take considerable research to design, however. Even though Cavette's invention is an easily manufactured system of straps, metal loops, and fasteners, it required time and effort for him to design it so that it could be manufactured profitably on a large scale. He made his initial prototypes by hiring a local seamstress with a heavy-duty sewing machine and providing her with supplies and items from a hardware store. Using this approach, the materials and labor involved to make each unit cost about three times as much as Cavette thought consumers would pay—nowhere near the profit margin that he hoped to achieve. And he still had to contend with developing a system of promoting the product, taking orders, and fulfilling them. Cavette knew he could bring

costs down and also create a better-looking product if he had access to the right machinery and the ability to buy materials in bulk.

Friends and family offered investments to help him manufacture the product himself, but, like most inventors, he was concerned that failure of his product to take off could damage personal relationships as well as professional ties. Furthermore, he feared that even if the product had meteoric success, it may not have the longevity in the marketplace that would justify planning a future around the product.

The simplest invention can take considerable research in order to design it to be manufactured profitably on a large scale.

Cavette also didn't know much about packaging, distribution, and other links along the supply chain. He was able to learn all this, as other inventors have done before him, but learning curves inevitably involve trial and error, and in this case every error came directly out of Cavette's pockets. Moreover, he genuinely liked being a firefighter, and had no interest in quitting his job and losing his benefits. His new product had the potential to be a profitable product, but it probably wouldn't make enough to live on.

As of the writing of this book, our firefighter inventor is still deciding whether to manufacture his invention himself or license it to another company. He is continuing to gather data on his target market and research a way to bring the manufacturing costs down, and he is keeping his options open, as we advise you to do. With three small children and a stay-at-home wife, however, Cavette views *eventual* licensing as a good way to realize a healthy return on his initial investment without turning his life upside down.

What About "Angel" Investors, Grants, and Free Money?

Some of you may be shaking your heads, saying, "Rather than limiting his profit potential to a 3 to 5 percent royalty, Cavette should just find some investors for his great idea and hire people to do the marketing." The fact is that the odds of getting perfect strangers to give you money to develop a consumer product you are not planning to market yourself are slim to none. If you are a first-time inventor with a consumer product, your best source of investment capital (as discussed in Chapter 7) will be your friends and family. They know you, they believe in you, and they will be the most flexible and understanding when it comes to repayment terms.

Banks are another good place to look for start-up funds if you have the borrowing base to allow for this. Banks will lend you the value of what you already own. They provide liquidity, allowing you to borrow

money against assets that you have but are not interested in selling. These assets include home equity, stocks and bonds, and jewelry. A third source of investment funds could be an *angel investor*. "Angels" are high-net-worth individuals (or groups) who are interested in investing their money in higher-risk deals. It's up to you to persuade them that they can 1) safely risk their funds with your product and 2) realize a high-enough rate of return to justify that risk.

Organized angel investor groups pool their resources to invest in promising start-up companies and focus on those in high-level emerging industries. The groups seek out meteoric companies to invest in that are run by qualified, committed management teams with the potential to bring multiple products to market. They usually invest in companies with intellectual-property portfolios, rather than in an individual patent that has no company or management team built around it.

John Abdo learned firsthand that persistence and passion can pay. Abdo had created the AB-DOer, a midsection aerobic machine that allowed him to exercise despite his chronic back pain. In attempts to raise capital, he had to risk almost everything. With a $100,000 debt and repeated opposition from venture capitalists, he was near bankruptcy, and things looked grim. Fortunately for Abdo, he had a successful syndicated TV series, *Training & Nutrition 2000 (TN2000)*, airing at the time. The show had caught the interest

Thanks to funding from an angel investor, John Abdo's AB-DOer was brought to market, earning hundrds of millions of dollars.

of an angel investor who was willing to take the risk. The fan agreed to fund the product development in exchange for a percentage of Abdo's business. With funding, Abdo, a passionate and persistent inventor, went on to much success with AB-DOer sales in the hundreds of millions of dollars!

It's unlikely that you'll find state, federal, or municipal funding to help you bring a product to market. Government funding is usually limited to investments in high-level technologies that provide a national benefit (such as the development of alternative energy or devices to reduce bioterrorism threats). Public funding is reserved for situations in which there is a public benefit to government for "outsourcing" scientific research to the private sector. There are some government grants, such as Small Business Innovation Research (SBIR) and Small Business Technology Transfer (STTR), that provide research-and-development grants and contracts to qualified small businesses to pursue research for the government on a specific, defined problem. Although these grants do not require repayment, they do require a lengthy application process and a credible organizational team to qualify. You can find more information at www.sba.gov/SBIR.

There are also Small Business Administration (SBA) loan programs, which help guarantee loans made by banks to small businesses. In order to qualify, you must make the commitment to start a business. You must pay back these loans, and the interest rates are comparable to commercial rates. In order to secure them, you are expected to put up collateral to guarantee the loan. The role of the SBA is to provide a coguarantee to the bank for loans made to companies that have collateral but no established credit.

In short, there is virtually no "free money" from taxpayers to bring your product to market. If you are light on funds, risk averse, or do not want to make a long-term commitment to starting a new business, licensing may be your best option.

Understanding the Legal Significance of a License

Although you may often hear people talk about "licensing an invention," this is technically incorrect—there's no such thing as an invention license. What they are really referring to is a license of the intellectual property (patent, trademark, copyright, trade secret) rights. We can't say it enough times: As a patent holder, you have the exclusive right to profit from the technology covered by the claims in your patent and to prevent others from competing with you.

Obtaining patent protection for your invention is essential in your ability to obtain a license. Without a patent application on file, you don't have any rights of value to license. In other words, your license agreement won't be worth a dime to the licensee without a patent, patent application pending, or other intellectual property.

The license agreement itself is actually a type of contract in which you "rent out" your patent rights in exchange for a royalty. A license is generally for a defined period and documents specific percentages or amounts to be paid to the licensor (you).

Another type of an agreement, called an assignment, is closely related to a license; sometimes the two kinds of contracts are confused. An *assignment* is an agreement under which you transfer your patent rights permanently to one party, rather than for a set period. Employees of companies are often required, as a condition of employment (and before actually inventing anything), to assign their patent rights to their employers. This is usually the case with any invention that was created with the employer's resources or within the scope of the employee's job responsibilities. The employer, as a result of the assignment contract, owns the right to enter into license agreements or otherwise profit from the patent.

Another important difference to point out between a license and an assignment is that with a license you may enter into simultaneous licenses with several parties. With an assignment, you transfer all your rights under the patent to the assignee. In contrast, a license may be nonexclusive, where you give several parties the right to produce and sell your product, or it may be exclusive, meaning you give only one party this right. Philo Farnsworth, the inventor of the television set, entered into non-exclusive license agreements with both British Gaumont and RCA (see page 146). That allowed both companies to simultaneously commercialize the television set and compete very intensely with each other. Generally, if someone wants an exclusive license from you, they'll have to pay more money, since it is a more valuable right. This should definitely be a bargaining point in the licensing negotiations.

Finding a "Licensing Partner"

Licensing can be an ideal partnership of creativity and capital in which both parties profit from developing and selling a single product. As an inventor, you can gain access to your partner's capital, manufacturing capabilities, and distribution outlets to bring your product to market or expand your existing market. From the company's perspective, you bring patent rights that can help the company get an edge on its competitors.

Companies are motivated to pursue licensing arrangements when individual inventors bring them good products that fill proven niche markets. Because the individual inventor holds the patent rights to the invention, the company has no choice but to negotiate with the inventor if the company wants the invention in its product line.

It's important to understand the market dynamic of the licensing arrangement. The licensing agreement is a type of joint venture in which both partners are hoping to make a profit from commercializing the new product, generally in proportion to their relative risks and investment.

In some cases a company must consider the risk of not licensing your patent. There is a chance you'll license it to their competitors or be able to develop it yourself and compete with them. The stronger your presentation, the more likely it is that they will perceive licensing your product as an opportunity they should not pass up.

Getting in Front of the Right Companies

If you have a good idea, you need to get it to the right place. Companies must assess how well your product fits into their existing product line. Can they use their existing manufacturing capabilities to produce it? Can they rely on their current outlets to carry the product? Will the product add to their brand identification? Will it increase their market share or will it cannibalize sales of their existing products?

Showing the company you have a good fit requires you to research companies that are in a strategic position to develop your product. Here are a few things to consider.

Look for companies that carry products similar to yours. If you wanted to buy your product, whom would you contact? What stores would you visit? What search terms would you enter on the Internet? Where else would you seek it out?

Find a personal connection. Do you know someone who works at a store on your list of preferred venues, a large customer of a particular store or chain, or perhaps a common interest? Do some networking to find a way to get a warm introduction rather than making a cold call.

FROM THE ENTREPRENEUR
DO YOUR HOMEWORK

Not only should you research your prospective licensing partner's product line, but you should also research the company's history and financial situation. You don't want to do business with a company that's on the verge of bankruptcy or one that is shredding documents. It's not just your money at stake here; it's your idea.

Send snail mail, not e-mail. Send the companies you identify a brief, professional letter. Find out the appropriate contact within the company (an owner or a vice president of sales is a good bet) and send it to them via hard copy. Never send an e-mail without following up with a hard copy, and never rely solely on filling out online invention submission forms (even if that is what the company website directs). Research the appropriate contacts and keep an organized list including their names and titles.

Include a marketing brochure, photos, and/or a data sheet. Provide information that briefly describes your patented product, and remember to indicate the "patent pending" status on these materials. Indicate that you have a pending or issued patent, but *do not* disclose your application number or provide a copy of the patent in these communications.

Follow up with a phone call. Wait two to three weeks or so, then follow up with a brief phone call. Try to speak directly to the person to whom you sent your initial letter, and avoid using voice mail.

Ask for a meeting. Your goal in writing these initial letters and making follow-up phone calls is to get the opportunity to hold a personal meeting to pitch your product. Be persistent in requesting this. A lukewarm response by telephone can turn into an enthusiastic review when people actually see your prototype, your thorough research, and whatever other visual aids you have to offer. It's always harder to say no to someone in person. In person, you have a greater ability to infect them with your passion and enthusiasm. Use this opportunity to focus on the value proposition and why it makes sense for them to work with you. Be confident but also realistic. The better prepared you are going in, the greater the chances of achieving success.

Preparing for the Pitch

The invention you are bringing to the table will be compared to other developing opportunities the company is probably considering. You may end up negotiating with the owner of the company, or you may meet with a member of the sales, product development, marketing, or even legal department. Regardless, these people are all accountable in some way for the profitability of the company, and you must convince them that your product will be profitable. Having as much data as possible to support this is essential, since companies can bring only a limited number of products to market at a time.

Presenting the Product

The more developed your idea, the more value it will have to a prospective licensee. Proven sales (even if the sales don't amount to a lot of units) can go a long way in demonstrating that the invention can be made market ready, that people are using it without encountering unexpected problems, and that your invention is something worth developing further.

Of course, for some inventions, it's just not possible for an inventor to come prepared with actual samples and proven sales. Companies realize this, and many alternatives are available to inventors when finished prototypes and test marketing are not possible. Don't sweat it. So long as a patent application is in place and you know your market cold, you have something you can sell. All you have to do is convey it in the time you're given (which will probably be somewhere between ten minutes and half an hour).

Since you won't have much time to make your pitch, keep in mind what is relevant to the company. In Chapter 2, we explained that companies want to know that you have identified an unmet need, how many potential buyers have that need, and why your product is the best option for serving that need.

We have seen inventors make successful presentations with sketches and mechanical drawings of their invention, or with very rough prototypes that illustrate merely how the invention will look (a "looks-like" prototype) or how it actually works (a "works-like" prototype). Homemade devices pieced together from hardware-store parts are not uncommon. Professional sketches and drawings are inexpensive and can bridge the aesthetic gap to build the vision.

If you have all this data in hand, be prepared to present it in the most appropriate format for the audience. Make eye contact and observe people's reactions as you are speaking to them and passing around your prototype. You should prepare your pitch carefully, and be able to introduce your invention in the first sixty seconds—also be prepared to deviate from your pitch if called upon to do so. Be sure to answer questions that are posed to you directly, rather than saying you'll "get to them."

It's important to be flexible and be responsive to the company's concerns. If they weren't interested in the first place, you wouldn't be there, so delivering a canned spiel that doesn't address what *they* need to know in order to move for-

---

Prepare your pitch carefully, and be able to introduce your invention in the first sixty seconds—also be prepared to deviate from your pitch if called upon to do so.

---

ward won't move *you* forward.

Whether you approach a company with a simple sketch, a prototype, or a test-marketed product in hand, the important thing to remember is that the company is looking for as much data as possible to evaluate the sales potential of your invention. It's your job to get in front of companies you think are the right fit and present evidence that allows them to evaluate your product against all the other products they are considering bringing to market.

Determining a Fair Royalty for Your Invention

Typical royalties range from 2 to 5 percent of revenues, but they can vary based on the product, the margins, and the industry.

Most consumer products fall in the 2 to 3 percent range. Software is an exception: It can be well over 10 percent. Generally, the amount of the royalty is based on the market potential of your idea and whether your concept is well developed.

Royalties are typically calculated from actual revenue, which is in constant flux. At the outset it's important to understand how the company calculates revenue attributable to your product and what costs and expenses go into this calculation. Prior to understanding this, you can't really have a meaningful conversation about royalty percentages because you don't know what you're getting a percentage of. Generally speaking, a royalty is paid on the gross sale received by the licensee from their customer (normally a retailer). The gross sale could be the wholesale price of the item (the sale price) minus shipping, discounts, returns, and allowances. This final number that is used to calculate the royalty could vary significantly depending on the amount of those deductions.

The following is a partial list of factors that may affect the amount of your royalty. During negotiations, you may consider pointing out some of these factors that add particular value to your product and justify the royalty you are requesting.

Initial investment and research: You are not selling only your patented invention, you are also selling the research that has gone into its development. The more research and knowledge you can associate with your patent rights, the more valuable it is. Be prepared to point out what you know, and all that your patent application covers, including potential future product lines that are protected under this patent or other patents.

Strength of patent protection: A good patent, when issued, will protect you not only from competitors copying your exact idea but also from copycat competition, design-around competition, and equivalents. It is important to have a good *utility patent* in place that clearly covers the commercially valuable aspects of your invention. (See Chapter 4 for more about patents.) In addition, you should have multiple patents to protect your idea. A patent that covers your idea as well as ways of getting around your idea has more value to a prospective licensee than one patent alone has.

Size of target market and strength of your market research: The larger your target market, the greater the number of units that may ultimately be sold. Always be prepared to credibly discuss, during the negotiating process, how you arrived at target-market numbers and why they are accurate. Sometimes you may be able to present a point that the company has not yet considered, such as a special insight about your market's preferences.

Risk of change in market versus time for development: A product offering a long market life means a longer time for the licensee to recoup their manufacturing investment and generate meaningful profits. Products take time to develop and get to market, and the longer they can be profitably sold once they arrive on the scene, the better. Costs amortized over a longer sales period mean greater profits, since development funds do not have to be shelled out up front. If the company does not have to recoup a large up-front development investment, you can argue that the company can afford to pay you higher royalties in the short term.

Availability of substitute or competing products: If you have a unique product that meets a need that is hard to fulfill with substitute products, you can point out how the licensee will lose a competitive edge by employing work-around solutions. A good example of a licensed product with this leverage is the "push" e-mail software component that allows a BlackBerry device to deliver messages instantly. For more than ten years, BlackBerry's competitors didn't have software components that did this, which allowed Research In Motion, Limited to maintain the market advantage.

Profit margin: The higher the profit margin on your product, the more profit the licensee will make and arguably can afford to share with you in the form of royalties. Software has a high profit margin because of the low production cost relative to the selling price to the consumer. While development costs are significant, typical licensing royalties for software are greater than 10 percent.

Product-line extension potential: If there is an opportunity to develop several profitable products to capitalize on the niche you've created, there is added value to entering into a licensing relationship with you and getting a foothold in the emerging market. Broad, well-drafted patent claims that potentially cover future product iterations or line extensions can be particularly valuable to both you and the licensee.

Licensing Land Mines for Inventors to Avoid

We've seen more than one inventor trip up by creating controversies and silly-string legal issues that undermine the goodwill and momentum of the licensing partnerships they've worked so long and so hard to create. It's always in your interest to have a good relationship with your licensing partner, and to keep the door open to future deals and modifications.

Here are some what-to-do and what-to-avoid tips to help you negotiate an agreement for a long-lasting partnership.

Negotiating your licensing agreement: Licensing agreements must strike a balance between protecting your rights and those of the licensee company. Both sides need to understand the terms of the agreement from their own perspective, as well as that of the other party. Inventors who don't understand the terms of the agreement, and expect the company to answer an excessive number of questions about interpretation during the negotiating process, can appear defensive or just unsophisticated in business matters. If you have gotten this far, hire an attorney to educate you about what you are signing, and don't strain the relationship by expecting the licensee to walk you through it.

Hiring an attorney: On the other hand, exercise caution in what attorney you hire. Be sure you get one who is well versed in negotiating on behalf of individual inventors. There are some assurances and protections that individual inventors need, and the attorney should be brief and friendly in procuring them on your behalf, always representing you as someone with whom the company will enjoy dealing over the long term. We've seen inventors lose licensing deals when their attorneys send demanding letters requesting extensive changes to things the licensor and licensee already agreed upon and don't matter much to the inventor in the first place. As an individual inventor, you are an unproven commodity, and so is your idea. Having your attorney negotiate from a spoiled celebrity perspective will just run up a big legal bill and get a long-term licensing relationship off to a bad start, if it starts at all.

The "greedy inventor" scenario: Understand that inventing a single, profitable product might not make you rich. At the very least, it should make you (as well as the licensee) a fair profit. Remember, a reasonable royalty for a patent license is 2 to 5 percent. If the company offers less than you expect, ask them why, and look at their calculations. Analyze what they provide, and provide them with appropriate data if you feel they have understated the market. They are about to become your partner. It's critical to both parties that the licensor/licensee relationship remains fair and that both parties communicate effectively about the product throughout the marketing process.

A shortsighted view of the licensor/licensee relationship: If you are bringing the company an invention in your field, and you feel you may have other inventions to bring them in the future, view the first licensing agreement as a door opener. If you feel this is the only invention you will have, decide carefully whether it is in your interest to negotiate a longer or shorter duration for the license based on the long-term sales outlook. Shorter agreements, if the company will accept them, may be to your advantage if you think proven sales will strengthen your negotiating position when the license comes up for renewal in the future.

Entering into an agreement before you assess all your options: Many inventors are so elated to find a licensing partner to develop their idea that they forget to check back with the other companies they've contacted. Always check with the other companies you've approached. Creating a sense of competition among them can strengthen your bargaining position immeasurably.

Forgetting to get a signed nondisclosure agreement: Remember that the company you're approaching as a partner today may turn out to be your competitor tomorrow. Be sure to get a nondisclosure agreement in place, even if you have a patented product, in order to protect yourself against attempts to design around your patent. These agreements may be hard to enforce as a practical matter, but they should at least set the tone that you expect data and specifications you provide to be returned to you and not reproduced by the company. Reputable companies will have no problem respecting your rights.

Disclosing too much too soon: Even if you have both a patent application and a nondisclosure agreement in place, there is nothing to prevent the company from using general market research you've disclosed to develop their own product line. Try to ascertain whether the company is serious about assessing your product and not stringing you along to find out how much you know about your market. If you get a sense that this may be happening, pull back. Assess the situation casually by asking

the company about its history in dealing with individual inventors. Legitimate companies will address your concerns in a frank, nondefensive conversation describing how they've approached and developed outside ideas in the past.

Being overly aggressive about confidentiality: Companies have lots of ideas to assess, and they have limited time. They need a reasonable amount of information to assess your idea. If you nitpick your nondisclosure agreement, you will come off as difficult to deal with from the start, and worse yet, the company may view you as litigious (someone who will be prone to suing them). Before you've even gotten through the door, they may feel it is in their best interest to simply move on. The best approach is to get a basic agreement in place, then disclose information appropriately, based on each stage of the negotiations. Think ahead of time what you need to disclose to make your case in the initial pitch, and what you don't. Always handle requests for information diplomatically rather than defensively.

Providing a copy of your patent application before it becomes public (published on the USPTO website): Until you have a licensing agreement or a *letter of intent* (a written statement of the intent to enter into a formal agreement) from a licensee in hand, there is no reason to show them a copy of the pending patent application. Pending applications are not published on the

United States Patent Office website prior to eighteen months of filing. This period of secrecy gives both you and the company a competitive edge in developing the product. Companies may not want to risk bringing a product to market if their competitors have already had the opportunity to review your unpublished patent to figure out how to design around your claims, make your product, and use it. Also, don't give them your patent application number; that would give them the ability to directly communicate with the patent examiner during that initial eighteen-month period (during which time they could perhaps present arguments as to why your application should be denied). Your patent attorney is a good resource to whom you may direct questions about your patent in the early stages of your negotiations.

Giving up rights you need to keep: Don't enter into an exclusive agreement unless you are compensated for doing so, and don't enter into a long-term agreement for a product without annual *performance metrics* (minimum performance requirements to maintain exclusivity) and minimum guaranteed royalties. This is critical and will prevent the ability of the company to "bury" your invention without ever bringing it to market or paying your royalties.

Creating your own legal conflicts: You can license the rights that you have, but it's important to know that all sorts of seemingly unrelated legal agreements can affect your right to enter into a license. For example, corporate shareholder agreements, partnership agreements, employment agreements, agreements giving creditors an interest in your assets, divorce decrees, and marital property agreements can all affect your patent rights. Don't enter into these types of agreements without checking with your patent attorney about how they can impact your ownership in your intellectual-property rights and your right to license them to others in the future.

Overlooking improvements that will occur over the life of the license: When negotiating the agreement, make sure that any improvements that you or the licensee make to the product are also covered under the royalty agreement. You also want to make sure that these improvements, if patentable, are considered to be derivative intellectual property and are assigned to you as the inventor to be protected as part of your intellectual-property portfolio. For example, if either party comes up with improvements or additional products based on your initial invention, you'll want to make sure your license adequately addresses your entitlement to royalties on those products as well.

Upfront, or advance, payment: Receiving an upfront payment from the licensee upon execution of the license is ideal. The payment can help you, the inventor, to recoup some of the development costs incurred. There may be reluctance on the part of the licensee to put up dollars before sales, so you should be prepared to support your position by explaining the significant investment already made on your part, and as a result, the savings the licensee has realized. In addition, an upfront payment shows commitment on the part of the licensee since they now have "skin in the game." One way to counter reluctance on their part is to treat the upfront payment as an "advance on royalties." An advance is just a prepayment of money they will eventually owe you anyway.

Most importantly however, you need to determine how far to push this and whether this is a "deal killer" if you cannot agree.

Setting performance metrics: Any license agreement should set out minimum expectations and performance requirements. This could include minimum royalty payments that are required regardless of actual sales activity. Setting these minimums prevents a licensee from just sitting on your idea. You should also use these to enable you to exit from an agreement if the licensee does not live up to their obligations.

Using an agent: We strongly recommend that you try to license your invention yourself. First, you are more knowledgeable about the product than anyone else is. Second, you are more passionate about the product than someone you hire. Most important, you

FROM THE LAWYER
PROFIT PROJECTIONS

At the initial stages of negotiating a licensing agreement, the inventor and licensee (manufacturer or distributor) review the projections of profits to be made and come to a sort of shared vision of the market. Remember, the manufacturer is taking a financial risk in deciding whether to market your product at all. You, as the inventor, should work to convince the prospective licensee that your economic interests are aligned and that this is too good an opportunity for the licensee to pass up. Keep it amicable—bringing a visible, verbal lawyer in too early can send the wrong message and play into something we call the "greedy inventor" scenario, which can kill a deal.

have the most to gain from the success of the product. It will require time, and you may face rejection, but the rewards are certainly worth it. The invention industry is filled with companies and individuals who advertise their abilities to license or promote your idea. Understand that many of these companies fail to deliver on their claims and are more interested in the upfront fees they receive than the royalties they can generate. Finding a reputable one is a challenge and requires a great deal of diligence. If a company asks for upfront money, that is a red flag. On the other hand, if they have a proven track record, are willing to be compensated based on performance, and allow you to exit from the relationship if they don't generate results, then using their services may be worth exploring further.

CHAPTER 7

Money Matters

Finding the Funds to Bring a New Product to Market

Graduate students Larry Page and Sergey Brin didn't exactly hit it off when they first met at Stanford in 1995. Sergey, then twenty-three, was assigned to show the twenty-four-year-old Larry around the campus on a weekend visit. According to Google legend, the two computer geeks argued about pretty much everything until they realized they agreed on one critically important point—a shared vision of solving the big issue in computing at the time: the need to sort relevant information from a massive set of data.

In three years, Page and Brin sold their possessions, maxed out their credit cards, and dropped out of Stanford. But they bought a terabyte of memory and created Google.

Within the year, operating from the college campus, Page and Brin had begun collaborating on a search engine called BackRub, named for its ability to analyze the "back links" that point a Web user to a particular website. Suffering from the lack of cash that nearly all students face, Page began directing his creative efforts to building a frugal but credible working prototype. He used low-end, secondhand PCs to test their method instead of fast, expensive ones, and "borrowed" computers from the school. The two sold personal possessions and used credit cards when they could get them.

Before long their new method of analyzing website links was working amazingly well and was spreading around campus. Page and Brin *knew* they had a product the market would embrace; the challenge was to find the funds to deliver it. They began talking to friends, family, and anyone else who might appreciate the opportunity. But it was still too early in the process, and no one seemed to grasp the potential.

In 1998 Page and Brin, now in PhD programs, were perfecting their technology, a search engine they called Google, and showing it to an expanding network of supporters. During this time they

bought a terabyte (a very significant amount of computer storage at the time) at bargain prices and set up Google's data center in Page's dorm room. Meanwhile, Brin continued to work on contacts and potential licensing partners, asking if they were interested in a search technology better than any currently available. The companies were not. Despite the dot-com frenzy, the better-established companies they approached did not think it was worth introducing the new technology to the marketplace; nor were they sufficiently worried about the competition Google might present to want to bring them in as licensees.

One person Brin contacted was Yahoo! founder David Filo. Filo agreed that Google's technology was potentially superior. But from a business standpoint, Filo felt that this upstart posed little risk to Yahoo! He told Page and Brin to come back when the technology was "developed and scalable." It looked as if Google would never see the light of day.

Finally, Page and Brin decided to market their invention themselves. They'd need to move out of the dorms (they'd maxed out their credit cards by this time). So they wrote up a business plan, dropped out of graduate school, and went looking for an angel investor. They found one close to campus, through a faculty member.

Andy Bechtolsheim, one of the founders of Sun Microsystems, took one look at the boys' demo of Google and knew what he was seeing. Brin later recounted, "We met him very early one morning on the porch of a Stanford faculty member's home in Palo Alto. We gave him a quick demo. He had to run off somewhere, so he said, 'Instead of us discussing all the details, why don't I just write you a check?' It was made out to Google, Inc., and was for $100,000."

Page and Brin weren't even incorporated yet, so they couldn't exactly deposit the check immediately. It had to lie in Page's desk drawer for a very long week while he and Brin set up an actual corporation. Once the first angel was on board, family, friends, and acquaintances all began to pick up their checkbooks for the duo as well. After three long years of conceptualization and networking, Google, Inc., opened its doors, having raised an initial investment of almost $1 million.

Starting Small . . . Realistically Assessing Your Financial Options

If you are a first-time inventor operating out of your home or even dorm room, the odds of getting outsiders to give you money for your product are slim to none. First and foremost, you must prove the product's market potential. In the early stages, successfully getting others to invest in your idea usually involves what's called bootstrapping. Bootstrapping, as mentioned before, means initially getting into the market using your own limited funds or, better yet, a creative strategy that doesn't require outside investors. Almost all successful inventors start out by doing some sort of bootstrapping—Page and Brin did, and most likely you will as well.

Bootstrapping Your Invention to Success

To bootstrap your invention is to fund it yourself or to find other ways to avoid using funds from outside investors. The term "bootstrapping" is, as previously noted, often attributed to American author Horatio Alger and his "rags to riches" dime novels of the 1880s. Independent inventors, like the characters in Alger's stories, have a number of resources at

Bootstrapping 101

It is amazing how many companies with global impact are launched from college dorm rooms. Google, Inc., and Dell Computers, Inc., were started while their founders were still in college.

College is a great time to start learning about launching a business for a lot of reasons. Your cost of living is low, and you have to do everything with limited funds because money is already tight. You have no kids to support, no retirement looming on the horizon. In general, the risks are much lower (if you fail, you can always graduate and get a "real" job), and the college environment provides a fertile ground to grow a business.

FROM THE ENTREPRENEUR
PUT YOUR EGGS IN DIFFERENT BASKETS

Funding the creative process all on your own is wonderful because you don't end up owing anyone after all is said and done. That said, the last thing you want is to invest all your savings only for your invention never to see the light of day. In the inventing world, don't be afraid to ask for outside help. Sometimes there's no way around it.

their fingertips to help propel themselves to wealth and success by their own hard work. Relying on private equity, venture capital, and other outside investors can cost 20 to 50 percent of the equity of your company, or more, and is likely to require you to form a board of directors on which outsiders will sit. Although self-funding means assuming the risk of success or failure on your own, it gives you complete freedom to develop your invention without being bound to anyone. Here are some common techniques independent inventors have used to bring their inventions to market.

Using Savings, Credit Cards, Retirement Plans

Most investors ask why they should risk their money if you don't believe in the concept enough to invest some of your own. As one angel investor said, "I assume there is something wrong with an idea if an inventor can't raise the first $10,000 to $50,000 before he comes to me." Although investors may require you to put your money where your mouth is, think twice about the point at which you do so. Developing any new technology or product is fraught with risk. If you and your spouse cash in your IRAs close to retirement, you must make sure that you can weather the potential storm. The odds are mightily against you until the point at which you can prove the market is actually (not hypothetically) embracing your product at a price that makes it profitable for you to produce it. Sound market analysis, surveys, and testing are all techniques we discuss in Chapter 2 to help you assess the risk of entering the market using a nest egg.

Many successful inventors get their concept launched by using their credit cards. While this is a tempting strategy, you must exercise extreme caution in deciding to use it. Most credit cards carry a very high interest rate for cash advances or for revolving balances. This can be an expensive "loan" if it is not paid back, and will have a detrimental impact on your overall credit. On the other hand, if you are comfortable with the risk and the numbers support it, credit cards are a quick and easy way to access capital.

Leveraging a Purchase Order to Pay Production Costs

You *can* realize a healthy profit without risking your own funds. Using a purchase order as leverage is a solid, realistic, and workable strategy, and if you have the charisma and tenacity to pull it off (which you probably do), you are very likely to succeed in the marketplace. Purchase orders from well-known and reputable retailers and distributors can be used as an asset from which to borrow. There are many banks, private investors, and even manufacturers who will use a purchase order (which constitutes a promise to pay) as collateral to advance funds for manufacturing.

Say, for example, a large retailer issues you a purchase order for $1 million. You will probably need at least $500,000 to produce the product and ship it to the retailer before you can get paid. If you go to a bank and use the $1 million purchase order as collateral, the bank might be willing to lend you the $500,000, knowing they will get paid with the proceeds. In many cases, the bank will have the receivable (amount owed by the customer) assigned to them: They collect on the customer's invoice and remit the difference to you.

Another way to get your business off the ground is to use your manufacturer to finance the order. Barbara Carey, a successful inventor, explains how she started out on the verge of bankruptcy, after an early failed venture, and rebounded with this strategy. Carey had an idea for a low-cost Halloween mask inspired by some masks she'd seen in Mexico made of unique materials that were perfectly priced, positioned, and packaged for Kmart. She convinced a manufacturer to provide samples at no cost based on the market potential of the product, and to accept payment

in ninety days, assuring him she'd be back with a large purchase order. The manufacturer believed in the product and the profit potential, and in return, Barbara agreed to purchase exclusively from them. It was a win-win situation. Shortly after her meeting with the manufacturer, she pitched the product to Kmart and got a purchase order that promised payment thirty days after delivery, which gave her plenty of time to pay the manufacturer within the agreed-upon ninety days. Carey earned a six-figure profit without having to shell out her own funds for the inventory.

Her formula debunks the myth that "it takes money to make money." Carey cautions, however, that she almost never goes into production until she has a firm purchase order commitment, and she pays the manufacturer in *less* than the time allotted in order to maintain her reputation for keeping her word.

To benefit from a strategy like Carey's, you need to:

■ **Locate a qualified manufacturer.** Ask pointed questions about your manufacturer's track record to ensure that they can and will deliver as promised. If they can't do what's promised, the payment terms become irrelevant, because you will have risked, if not ruined, your relationship with your customer. Whether you are working with a domestic local manufacturer or one overseas, it is important to meet with them personally and feel comfortable

Barbara Carey convinced her manufacturer to finance making the samples of her Halloween mask with the promise of a large purchase order.

with their capabilities. If overseas travel is not possible, then make sure you're convinced that your sourcing agent has the ability to perform on your behalf.

■ **Negotiate favorable terms with the manufacturer.** Try to arrange terms with the manufacturer that allow you enough time (sixty to ninety days) from the date of delivery to collect from your customer, leaving a margin of time for small setbacks and delays that may arise. Also, negotiate with the manufacturer to procure samples at little or no cost, pointing out specific factors that indicate to them that by helping you at this stage, they are working at developing a profitable relationship.

■ **Procure a purchase order.** Once you've lined up your sourcing and secured the best quality and terms, you need to pitch your product to retailers. The goal is to procure one or more solid purchase orders from a major retailer who is willing to pay you within thirty to sixty days after delivery. These are typical terms.

■ **Keep your promises.** Deliver the product and pay the manufacturer on time. It is important to do so in order to maintain the relationships that will allow you to produce and pitch future products and build credibility in approaching other companies. (And you will have profited nicely with a minimal cash outlay—a process that just might be worth repeating!)

Are there any downsides to working exclusively with one factory? Of course there are. You may be locked into the relationship with your partner, which prevents you from pursuing others. The manufacturer is, quite fairly, charging you a premium for financing your deal. You also assume the risk of paying the manufacturer regardless of whether your customer pays you. Though this risk is generally small, even major retailers have been known to go belly up.

Funding with a Factor

Working with a *factor*, a person or bank who looks at the value of your receivables rather than your creditworthiness, is another strategy that allows you to profit from a product without paying upfront. It's structured so that you use an account receivable (in other words, money someone owes you) to finance a transaction. Factoring is not a loan. Rather, the factor is buying an asset from you, which is your invoice (account receivable) or an issued purchase order.

Factoring is an attractive way of raising cash for small, innovative, fast-growing firms. It involves advancing funds on a purchase order or advancing payment on an invoice prior to payment by the customer. One method is preferable if you need the money required to produce your product, and the other if you need the cash flow and cannot wait the typical thirty to sixty days a customer takes to pay your invoice. The factor advances a percentage up-front and then pays the balance, less their fee, upon payment by your customer. Factoring a purchase order is more expensive than factoring an invoice because the factor's money is outstanding for a longer time. As Dylan Morgan, a principle at Prairie Business Credit factoring firm, explains, "It is an expensive form of financing, running as much as five percent a month. We like to see companies use us to get started, not as a long-term solution." He explains that factoring is a good solution for high-growth companies that have existing receivables with creditworthy customers and are looking to finance production on successively larger purchase orders.

Morgan takes a personal interest in seeing companies succeed, but he insists that factors must be exceedingly cautious about the risks they assume. Here are some of the issues Morgan's company considers in deciding whether to extend credit to an inventor.

■ **Gross margin:** *Gross margin* refers to the amount of profit you make on a purchase order after you subtract the costs of producing it. Generally, it is expressed as a percentage according to the following equation:

$$\frac{\text{gross margin}}{=}$$
$$\frac{\text{revenue} - \text{cost of goods sold (COGS)}}{\text{revenue}}$$

Morgan typically looks for a gross margin of 30 percent or more, although there are exceptions. If the gross margin is too small, the inventor ends up paying the factor too much and will not profit, and could also end up losing money.

■ **Credits that can be taken against your purchase order:** Look at the terms of the purchase order to make sure it's an actual purchase order and not just a consignment arrangement with a "guaranteed sale" provision. The latter would mean that the customer could return any unsold merchandise to you. Regardless of whether the document is labeled as a purchase order or not, if the company can return unsold merchandise, they aren't really buying it from you. In fact, you won't know if you've earned any money, or how much, until the time has passed for returning the goods to you.

■ **Ability of your company to meet the purchase order terms:** Unrealistic delivery schedules are a red flag. A good factor will make sure your company can meet projected delivery dates and other production requirements.

■ **Strength of the company issuing the purchase order:** Most purchase orders from major retailers are acceptable, if they contain the right terms. There are, of course, exceptions—some major retailers have a poor history of payment or are themselves experiencing financial difficulties. This can be hard to know, but a good factor will look into it. Also, you should carefully research all media and publicly reported information about retailers with whom you hope to do significant business, and be alert to any reports of falling earnings or financial trouble.

■ **Likelihood that the product will sell:** If a company places a large order for a product that doesn't generally appeal to its walk-in customers, it's worth looking into. A hardware store placing a large order for apparel, for example, will raise questions that need to be answered before lending against the purchase order.

You should also talk with your factor about collections and guaranteeing receivables. Factors generally advance 75 to 90 percent of the purchase order value and return the balance (less their fees) after they collect from your customer. Some factors will hold you responsible in the event the customer never pays; others guarantee payment and hold you harmless. Each approach entails a different percent fee, so make sure you look at your options.

Looking at Letters of Credit

A common form of paying an overseas factory involves issuing a *letter of credit,* a guarantee of payment to a manufacturer or supplier made by your bank to their financial institution. It is used to pay the production costs either domestically or overseas. Domestic letters of credit are used to guarantee that you will pay your manufacturer, within the terms, and that in the event you do not, the manufacturer can draw on the letter of credit to get payment. As an example, you obtain net thirty days (N/30) terms from a new manu-

facturer who requires, in order to get these N/30 terms, a letter of credit. As long as you pay your vendor within thirty days, the letter of credit is not drawn upon. However, if you fail to pay your supplier on time, they can present the letter of credit to your bank for payment. In order to qualify for a letter of credit, you need to put up the necessary funds or need to have the borrowing ability with the bank.

Foreign letters of credit are somewhat different in that they are actually used to pay the factory, not to guarantee that you will pay them separately from the letter of credit. To obtain the letter of credit, you put up the funds with the bank, and they issue an international letter of credit that is payable to the factory's financial institution upon presentation of a bill of lading for the merchandise. What this means is that once the factory produces your product conforming to your purchase order to them (quantity), and they deliver the order to the shipper (truck or ocean-going vessel), they have met the terms of your purchase order and so can present the letter of credit for payment. Your bank then releases payment to their bank. *You* now own the goods. A bank will generally go through its underwriting (qualification) process in issuing a letter of credit and review your company's performance history over several years. If you need an international letter of credit, make sure your bank has the capability. Many local banks can issue only domestic letters of credit.

CASE STUDY

Some Famous Bootstrappers: Analyzing How They Did It

Cofounders of Apple Computer, Inc. (now Apple, Inc.), Steve Jobs and Steve Wozniak, both began their careers by dropping out of college. Instead of attending classes, they began selling their own custom-built devices to allow people to make long-distance phone calls. In 1976 they sold their personal possessions, a van, and two calculators to raise $1,300 to embark on the venture now known as Apple, Inc. As of 2008, the company is worth more than $158 billion.

Countless inventors have bootstrapped their inventions to success, without initial outside funding. Here are a few more business-minded bootstrappers you may recognize, and a quick rundown of their strategies.

- **Borrowing from a friend:** Calvin Klein's childhood friend loaned him the $10,000 he needed for a dingy showroom to exhibit a small line of samples in New York City's fashion district. When the vice president of Bonwit Teller made a wrong turn while he was out walking one day, he happened to see Klein's samples. The VP had Klein bring his samples to the president's office, allowing Klein to bypass the buyer.
- **Mortgaging the house:** Despite numerous warnings of the competitiveness of the food industry and the fact that she had no experience in the business, Debbi Fields was determined to go ahead with her vision of a cookie empire. She and her husband went to their home mortgage lender and put up the house as collateral. Millions upon millions of Mrs. Fields baked goods later, Fields wrote in her book that their bank "trusted us, not cookies."

Steve Jobs and Steve Wozniak (in 1984 with then president and CEO John Sculley, center) bootstrapped their way to founding Apple, Inc.

- **Getting a loan from a credit union:** Domino's pizza, which has become famous for its fast pizza delivery and many locations, was started when two brothers, Tom and Jim Monaghan, obtained a $900 loan from the Post Office Credit Union.
- **Starting small and simple:** Ben Cohen and Jerry Greenfield started Ben & Jerry's ice cream with relatively modest expectations. They completed a correspondence course in ice cream making from Pennsylvania State University and, finding it to their liking, they borrowed $4,000 and compiled their savings to come up with a total of $12,000 to turn their hobby into a business.

Tips from Inventors Who Did It Their Own Way

Bootstrapping strategies take creativity and tenacity. Seasoned inventors often aggressively control costs, convincing manufacturers to extend payment terms that allow them to get paid by their customers first, enforcing strict collection policies, and operating virtual businesses from their homes.

■ **Operate a virtual business (from your home or dorm room).** Technology now allows just about anyone to operate a credible home-based business with minimal support staff. From your computer, you can communicate easily with manufacturers and customers. Operating your business from home could save you a fortune in the early stages, until your staffing needs force you out of the house. Operating at home eliminates rent, commute time you could devote to your business, extra phone lines, and more—not to mention a nice tax deduction.

■ **Accept credit cards and reduce receivables.** Let someone else assume the risk of receivables and the cost of floating "a check in the mail" by accepting (and insisting upon) credit card payments. You get paid and have access to the cash almost immediately. You can accept credit cards online, over the phone, or by fax. Again, technology makes credit card payments easy to process over the phone, and the fee for accepting credit cards is very rea-

sonable, typically between 1.5 and 3.0 percent of the transaction.

■ **Use drop-shipping services.** These services can save you the up-front investment of rent and staff during your startup stages. Drop-shipping, or fulfillment, is a service offered by a vendor that allows you to transmit your customers' orders to them. They then directly ship the product out with your label on it. Drop-shipping saves you the cost of warehousing your inventory and setting up your own shipping capabilities.

■ **Leveraging with licensing.** Licensing, as discussed in Chapter 6, can be the ultimate bootstrapping strategy. If you are short on funds, have a great product, and are willing to fork over a large share of the profits for someone else to manufacture and distribute the product for you, then licensing may be the best option.

Contacting People You Already Know for Funds

If you have a good idea, at some point others will be willing to help you. And those people are most likely to be people you know, who believe in you and your idea. After you've exhausted as much of your own funds as you can reasonably afford, and the idea is being well-received in the marketplace, it is time to begin relentlessly making contacts. If friends and family

Is Drop-Shipping Right for You?

Order fulfillment, or drop-shipping, is a great option if you have an inventory of finished manufactured products to sell, but no warehouse or store in which you can keep that inventory. It's the method of choice for those inventors who sell their products online, and may be the answer for some website owners who want to become stockless retailers. This set-up allows them to keep no stock themselves. Instead they promote products on their website and get a fulfillment company to pick, pack, and ship these products for them.

Here are the steps involved in drop-shipping.

- First, you open an Internet Store with a shopping cart and accept credit cards on your site, or have a Web development company help you create one using a number of tools that are commercially available.
- Search the Internet to locate a company ("drop-shipper") with fulfillment services or a warehouse that will accept the delivery of the packaged products you want to sell.
- Set up an account with the drop-shipper of your choice to keep track of your customer orders and inventory, and agree on a fee you will pay for the drop-shipping service.
- When a customer buys a product from your online site, they will pay you by credit card, or cash, using a service like PayPal.
- After they pay you, transmit the order to the drop-shipper. The order will include the customer's name and address and shipping information.
- The drop-shipper ships the product to your customer with your business name and label on the package, so it looks like it came from your place of business.

don't have the funds to invest personally, they can introduce you to people who do. According to attorney Richard Kranitz, who has assisted in procuring funding for many ventures, "it is the ability and willingness to make these calls and contacts, and to persist in widening the circle" that determines who gets funded and who doesn't.

If you're looking for more funding than you can raise through friends and family and have already looked at traditional bank loans, your next best source of investment capital is private equity, typically high net worth individuals with whom you have personal contact. You must provide enough data to per-

suade them that they can safely risk their funds with you and realize a high enough rate of return to justify the risk. The information they will find persuasive depends on the individual, but at a minimum they need to understand your product and see some reliable numbers and projections. If you have initial sales, or an impressive prototype, this may be what you want to show them. You should have something written, but make it short and factual. Most individual investors won't read or rely too much on a template business plan document.

As much as we encourage you to tap into your network of family and friends, and to give them the first crack at a great opportunity, we also caution you that more than one friendship (or family) has faced troubles over a business relationship. Here are a few general rules we recommend to manage expectations.

■ **Do not borrow prematurely.** Don't borrow money from friends and family until you are far enough along to be certain that you are offering them a solid opportunity. Put up your dollars first, as a sign of confidence and to avoid conflicts later.

■ **Identify what you are offering your investors.** You can offer a share of profits, equity in your new company, or debt that is paid off at a favorable interest rate. Generally, equity or a well-written debt instrument (loan agreement) is the most straightforward. Splitting profits can be difficult unless the expectations are to simply sell the invention and split the proceeds.

■ **Set up a repayment schedule.** No one should borrow or lend money without a written agreement spelling out repayment terms, even if those terms rely on event milestones rather than strict timelines. Managing initial expectations is critical and will help avoid problems later.

■ **Spell out how the money is going to be used.** Nothing is more disconcerting to an investor than learning their funds went toward your living expenses when they expected them to go toward equipment, inventory, or working capital. Have a plan, stick to it, and keep detailed records.

■ **Spell out the risk.** Specify what happens if you can't make a payment, and explain in writing the risks involved. This is an uncomfortable conversation, but in fairness to your investors, it is one you must have, and have before the investment is made.

Approaching Angel Investors

Private equity capital (or angel investors) fills the gap in start-up financing after "friends and family." Angel investors can be individuals or small groups made up of individuals who pool their resources. They tend to take equity in either promising young companies or high-level emerging companies that interest them. (If they made their money in the technology field, they're often able to spot the technology-related winners at the starting gate.) In order to protect their investment, the angels typically want to become involved in the management of your company in some capacity—by being on your board of directors, for example. In the best case scenario, the angels bring more than just money; they bring power, contacts, and credibility in the industry.

Most angel investors keep a low profile; they do not want to be hounded by hungry inventors. Thus, they can be hard to find. As with your friends-and-family networking, you'll have to be persistent in making contacts that will lead you to the right angel investors.

Angel investors will be looking at your team as much as your invention. They want to know that you and your team have what it takes to successfully sell your product. Most times, they desire companies with the potential to bring multiple products to market, not single products.

Approaching Venture Capital (VC) Firms

Unlike angel investors, who typically invest their own funds, venture capitalists professionally manage the pooled funds of others. Most traditional venture capital firms do not usually consider investments less than $1 million to $2 million. Like angels, VC firms speculate on certain high-risk businesses capable of producing a very high rate of return in a short period of time. They typically invest for periods of three to seven years and expect at least a 20 to 40 percent annual return on their investment, if not more. The reason VC firms look for such high returns is that many of their investments will fail to produce any returns; therefore, the investments that do yield positive returns must do so to support those that fail.

Doug Tucker, an attorney with the Milwaukee-based Quarles & Brady firm, has guided many companies through the venture capital process. With the number of requests venture capital firms receive for initial meetings and presentations of business plans, Tucker knows

that there are many unique ways for you to create a favorable impression with VCs. Likewise, there are many ways you can quickly lose their interest.

Tucker personally advises companies going for their initial round of venture capital not to make any one of the following ten deal-killing statements (otherwise known as "Ten Things Not to Tell a Venture Capitalist"):

1. "We have no competition." *Everyone* has competition. There is no faster way to communicate "we have no idea what we are doing" than to say this.

2. "We are partnering with so-and-so." (Microsoft, IBM, et cetera) when, in fact, all you are doing is purchasing products or services from them. These and other exaggerations (like, "We are the leading . . .") immediately call your credibility into question.

3. "If we only get 1 percent of the market . . ." If this forms the basis of whatever potential success story you are about to recite, you probably are not going to be receiving a term sheet anytime in the near future from a VC. Financial projections need to arrive at a well-reasoned result—there's no jumping to an easy conclusion.

4. "Our financial projections are conservative." Given the fact that less than 1 percent of all startups ever meet their initial financial projections, this sort of talk just makes you sound like every other amateur they hear from. Better to say "our projections are based on the following assumptions, but we have also run models demonstrating the level of sensitivity to a variety of factors." Adding that you would also be happy to e-mail your detailed projections in Excel format (and then actually doing it) may earn you bonus points.

5. "We aren't really looking for VC funding." This statement, frequently intended by entrepreneurs to imply a "hard to get" image, leaves a VC wondering why you are wasting your time (not to mention theirs) speaking to them.

6. "Will you sign a Nondisclosure Agreement a.k.a. Confidentiality Agreement)?" Venture capitalists will not sign NDAs unless there is a compelling reason (for them, not for you) to do so. This compelling reason usually comes *after* they have made a preliminary investigation into most of your business and technology and now need to see the "secret sauce" before funding the deal.

7. "I'll quit my full-time job when I get funded." If you have not fully committed your time, reputation, and finances to this project, why should a VC? Be able to enthusiastically, succinctly, and sincerely express your commitment to the project.

8. "The other VCs I have talked to have not understood what I have here." Most good VCs have a healthy respect for their industry colleagues and will be more inclined to believe that

their brethren understood all too well and turned a cold shoulder.

9. "I should remain CEO for the long haul because . . ." or **"I expect to maintain a controlling interest in this company because . . ."** Each sentiment shows an unwillingness to sacrifice control for the long-term best interests of investors.

10. "We've been too busy to write a business plan (or do financial projections)." Many VCs will tell you that they do not care that you have not prepared a business plan or financial projections. What they really mean is "I'll listen to you for a few minutes just in case you really do have the next big thing."

Getting VC funding for your invention is rare, and occurs only when the invention has the potential to be a huge success. This step in the process occurs after you have gone through the other sources of funding. Going to a VC before you have exhausted these others is a waste of time. But, Tucker advises, "there are exceptions to every rule."

Going Public

Every inventor looks forward to the day their invention brings them the riches they have dreamt of. After your product has gained market traction and you have created a successful company, you may find that additional financing is necessary to take your business to the next level. This method of raising capital involves offering stock in your venture to the public. The first time you sell stock to the public, the event is referred to as an Initial Public Offering (IPO). IPOs must comply with state and federal regulations and are typically brokered by an investment banker.

The primary advantage your business can gain by going public is access to capital from a far wider range of investors than it could otherwise reach through private solicitation. Being listed can provide an immediate infusion of capital to your company and, more significantly, your company may find it easier to obtain capital for future needs through new stock offerings. Investment capital (as opposed to debt) does not have to be repaid and doesn't require a company to pay interest. The only reward that investors seek is an appreciation of their investment, and possible dividends.

Additionally, going public may provide the business's founders and angel investors with liquidity, an opportunity to cash out on their investment. The shares of the founders and initial venture capital can be sold as part of an IPO, or on the open market some time after the IPO is completed.

Another advantage of going public is increased public awareness of the business itself. Usually a lot of publicity accompanies an IPO. However, a number of laws cover the disclosure of information during the IPO process, so business owners and managers must

be careful not to get caught up in the press interest.

Going public also allows a company the ability to use stock in creative incentive packages for management and employees. Offering shares of stock and stock options as part of compensation enables your business to attract better talent by providing them with an incentive to perform. Employees who become part owners through a stock plan are, at least in theory, motivated by sharing in the company's success.

Finally, an IPO provides a public means of valuation of a business. This means that it will be easier for the company to enter into mergers and acquisitions, because it can offer stock with an established market value.

Going public is not without its disadvantages. First, it is extraordinarily expensive. Legal and investment banking fees can consume 10 to 20 percent of the capital raised. Additionally, your company's management will be preoccupied with consulting investment bankers, attorneys, and accountants to make sure the IPO raises the anticipated revenue, and that the process does not run afoul of state and federal securities laws. Second, you contend with the loss of direct, ongoing control of your company. Federal law (specifically,

By going public, your business gains access to capital from a wider range of investors, but it can be very expensive.

the Sarbanes-Oxley Act) and listing requirements for stock exchanges generally require you to have an independent board of directors who do not have a financial stake in your company.

Getting Creative

We have talked about a number of traditional forms of raising money, but sometimes the nontraditional forms get overlooked. As an inventor, you need to get creative and find the money you need to get your idea to market. One way to creatively finance your idea is by raising the necessary funding from your customers.

By getting your customer to prepay for the order, you can use these dollars to produce the product for them. Why should they do this? Well, if they like your product, will benefit from it, and cannot get it anywhere else, there is a compelling reason to prepay. In addition, you might consider offering them a discount for the prepayment. Your margin will be less, but you didn't have to give up equity in your company to get it.

Enlisting Corporate Partners

Large corporations can also be excellent sources of relatively lower-cost capital, and may enter into deals with you to finance and purchase your invention rather than license it. Generally, to procure a deal like this, you must have the following:

■ **Expertise they can't get in-house:** Companies may be willing to fund you as an outside project, but only if they can't do it on their own with existing staff and resources.

■ **An in-house champion:** Finding someone in-house who is willing to go to bat for your invention is key; generally it takes an inside track or personal introduction.

■ **Time on your side:** Large organizations move slowly. The process of waiting for them to give your project a green light has been analogized as "dancing with elephants." Be patient.

Traditional Bank Loans and SBA Loans

Banks will lend you as much money as you already have. Sounds strange, but banks loan money to people who have assets that can serve as collateral for the loan. In essence, banks provide liquidity. They give you access to money without the need to sell your assets. Banks require collateral, which can be in the form of home equity, physical assets, stocks, et cetera. Many inventors have been known to lament that "they give you money only when you don't need it." Second mortgages are frequently used to convert the equity available in real estate into cash to fund your venture. Banks rarely will take stock or other equity in a start-up company, although they may insist upon taking a security interest in your business's equipment and receivables in case you default. Usually, if you're the owner of a start-up company, the bank will require you to personally guarantee the loan.

Small Business Administration (SBA) loans are available to guarantee loans when the business owners do not have sufficient collateral or credit history. These loans are very similar to traditional bank loans, and actually begin with discussions at the bank. The difference, however, is that an SBA-guaranteed loan involves the SBA guaranteeing a portion of the loan, which allows the bank to make the loan to you. While an SBA guarantee can rescue a borrower from lenders seeking high interest rates for high-risk loans, the government does not assume your debt. You still need to personally guarantee the loan, and if you default, both the SBA and the bank will come after you.

The Truth About Government Grants and "Free" Money

The U.S. Government invests about $70 billion annually to outsource needed research and development to the private sector and to various public institutions. A variety of programs are available to fund development of new technologies that benefit the public. If your product has military or environmental applications, and is something the government will pay to buy or develop directly, this may open some doors for you. These doors, however, can be hard to budge because of timing and bureaucracy. Nevertheless, you should assess for your-self whether you might meet the very stringent criteria to qualify for a government grant program, or whether you should partner with a business or institution that might be able to help you do so.

SMALL BUSINESS INNOVATION RESEARCH (SBIR GRANTS) AND SMALL BUSINESS TECHNOLOGY TRANSFER (STTR) PROGRAMS Federal agencies and departments spend $100 million annually on research and development of new technologies, such as bioterrorism threat defense methods, communications, and medical and military technologies. The federal government requires federal agencies to set aside 2.5 percent of their research budgets to contract with small, technology-based businesses.

Each government agency and department sets up and administers its own program in different ways. Some agencies solicit concepts that companies or inventors identify, others ask them to submit proposals addressing relevant, specific problems that government agencies are trying to solve. Phase I grants can be up to $100,000. If Phase I is successful, Phase II SBIR awards up to $750,000 in additional funds to help you finance commercialization of the invention.

These programs are very competitive. You must submit a very comprehensive proposal detailing your technology. The odds of winning a Phase I award are about one in ten, and they're about one in three for Phase II awards. To remain competitive, small businesses must successfully commercialize the results of the funded research. More information is available at www.sba.gov/SBIR.

ENERGY SAVING PROGRAMS The U.S. Department of Energy has had programs that provide financial and technical support to bring energy-saving inventions to market. These programs provide financial assistance: of $40,000 to $100,000. The amount awarded depends on the stage of development of the invention, and how well the inventor is able to convey the technology. This program is more accessible to independent inventors who are not scientists than the SBIR and STTR programs, but is still highly competitive.

ADVANCED TECHNOLOGY PROGRAMS These types of programs are intended to foster a sort of partnership between the U.S. government and private industry. They are generally administered by the National Institute of Standards and Technology (NIST). The purpose of the program is to speed up the development of high-risk technologies that promise significant commercial payoffs and widespread benefits for the economy as a whole. Applicant businesses must fund at least half the project costs, including all the overhead that is not directly associated with the process.

Without the necessary funding, your great idea will never be more than that . . . an idea. It requires a great deal of resources to go from an idea to a product that wil make you money. Understanding the financial risks is important, and it is critical that your investment in this process yields an economic return—to put it bluntly, it has to make a profit.

Regardless of how much or little you need, the amount does not matter as much as the return you will make on that investment. Building a compelling case for this return is a necessary requirement not only in raising money, but more importantly, in allowing you to determine whether it makes sense to even pursue funding in the first place.

If you have 1) a product that consumers want, at 2) a price that yields profit, and 3) there are enough consumers to 4) generate a meaningful profit, then funding will happen. You need to be diligent—don't give up.

CHAPTER 8

Protecting & Defending

Your Invention in a Global Marketplace

There's a saying that imitation is the sincerest form of flattery, but try telling that to Robert Kearns, inventor of the intermittent windshield wiper. At the time of his death, Kearns had received nearly $30 million in settlements from Ford and Chrysler for infringement of his patented intermittent windshield wiper, and numerous other suits were pending.

Kearns grew up in Gary, Indiana, near the giant Ford plant in River Rouge, Michigan, with a reverence for the automotive industry, and went on to become a highly respected professor

Although he was the first to invent intermittent wipers, Kearns (pictured here with his son) wasn't able to get to market before his competitors.

with a doctorate in engineering. Kearns got the idea for the intermittent windshield wiper on his wedding night in 1953, when a champagne cork struck him in the left eye. As he stood there blinking, he wondered if there might be a way to make windshield wipers that worked like an eyelid, moving at intervals that would allow for uninterrupted vision. Talk about a "Eureka!" moment.

After years of experimenting in his basement, he began applying for patents and approached the Ford Motor Company for possible licensing: He mounted the wipers on his 1962 Ford Galaxie and drove to Ford's headquarters. Ford's engineering team was amazed; legend has it that they swarmed into the demonstration room to see his invention. At one point they even ordered Kearns out of the room because they thought he might be activating the wipers with a button in his pocket.

There was no question that Kearns was the first to invent this device; Ford's engineers had been experimenting with a type of vacuum-operated wiper, but the intermittent wipers were totally novel. After the demonstration, however, Ford stopped answering Kearns's calls. There

was no offer to license. Over the next five years Kearns received numerous patents on various aspects of the invention, with the intention of bringing the wiper to the market on his own.

Unfortunately, Ford beat him to market, coming out with their own motor-driven intermittent wipers in 1969. Other major automobile manufacturers followed. Convinced that the invention had been swiped, Kearns's son bought an electric circuit for a Mercedes-Benz intermittent wiper, and as he expected, it was nearly identical to his father's invention. In 1978 the elder Kearns filed suit against Ford for patent infringement, seeking $141 million in damages; he brought suits against twenty-six companies altogether. Twelve years later Ford offered a multimillion dollar settlement, which Kearns refused. In July 1990 a federal jury ruled that Ford had unintentionally infringed on Kearns's patent and awarded him only $10.2 million.

Despite the money he had been awarded from settlements, Kearns was devastated that he would not be the sole producer of his invention.

Protecting Patent Rights in the Marketplace

If you develop a product that has the potential for great success, it's a good bet that others will try to copy it. While most patents are not infringed, understanding the potential risks of inventing is critical to assessing the value of the potential rewards. Patent litigation is a risk, no matter how remote.

The more successful your invention becomes, the more likely it is that others will try to design around it. Ideas with great profit potential give competitors major economic incentive to bring something similar to the market, even if that means infringing your patent rights. And big companies with deep pockets can afford the risk of being sued by individual inventors if they believe that the profits from their infringement will be worth more than a settlement. Being copied by others is outside your control; however, infringing someone else's patents is not. As an individual inventor (and especially with limited funds), you want to avoid inadvertently infringing others' patent rights by carefully designing around other prior art so your potential profits aren't spent on litigation fees.

A patent protects against more than just direct copying of the invention. It bars competitors from bringing functional equivalents to market. You can be sued and sue others even if the product is not a direct copy of a function and concept that is already patented. In the late 1970s the Polaroid Corporation received several patents (see page 190 for patent no. 2,435,720) for its instant imaging process, which develops photographs within minutes. Four years after the debut of Polaroid's instant imaging

camera, Kodak, a major competitor in the industry, introduced its own version. Kodak hoped that its process, which was based on a technology that functioned the same way with only slight modifications in the film development process, would be found distinct from what had been disclosed in the Polaroid patent.

Polaroid promptly filed suit, alleging that Kodak infringed its patents for the SX-70 camera. Kodak responded that the Polaroid patents were invalid and unenforceable and, in any event, were not infringed because there had been some modifications to the process. After seventy-five days, the court found that seven of ten of Polaroid's patents were infringed and awarded nearly $900 million in damages. Kodak lost on appeal and entered into a voluntary exchange program with customers, offering to swap its noninfringing camera for an estimated 16 million infringing models already in the market.

As the Polaroid suit aptly illustrates, you can *indirectly* infringe on a patent by making a product that has only trivial modifications and works pretty much the same as a competing product,

Before digital cameras, Polaroid was at the forefront of instant imaging photography.

FROM THE LAWYER
FOREIGN MARKETS

The U.S. is the largest market in the world, with a well-developed legal system of protecting your rights. Filing for foreign patent protection can be costly and should only be undertaken in countries where you feel you will actually have the potential to distribute your products.

Feb. 10, 1948.   E. H. LAND   2,435,720

APPARATUS FOR EXPOSING AND PROCESSING PHOTOGRAPHIC FILM

Filed Aug. 29, 1946   2 Sheets—Sheet 1

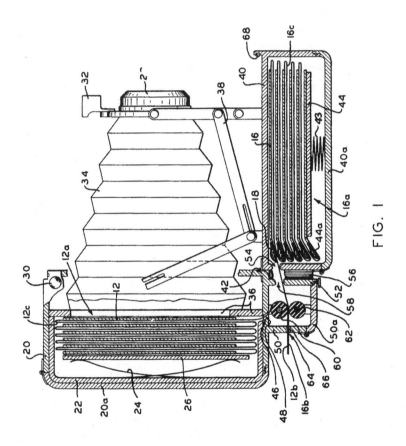

FIG. I

INVENTOR

Edwin H. Land
BY
Donald L. Brown
Attorney

Detailed drawings, such as this one in a Polaroid patent, can help protect your product from infringement.

whereby your consumers have no reason or ability to differentiate. If you "willfully" infringe a patent (i.e., flagrantly disregard the patent holder's rights), you can be liable for three times the amount of your competitor's lost profits or for other damages (treble damages).

As mentioned earlier, a patent gives you the right to exclude others from using, making, or selling your invention. It does not give you the right to sell your invention. This is especially important to understand if your invention is an improvement on someone else's patented product. You can patent an improvement to someone else's invention, but that patent does not give you the right to sell the improved product, just the right to prevent others from doing so. If you do improve someone else's patented invention and market or use the patented device (with your improvements) without permission of the patent holder, you may find yourself a defendant in a serious lawsuit for patent infringement. So, if your invention is based on an underlying patented product, you need to do a great deal of research to make sure you can safely bring it to market without infringing.

You *may* be able to patent your improvement under a separate application, but this does not give you rights to the original device. Your options consist of waiting for the original patent to expire, or trying to obtain permission or a licensing arrangement with the original patent holder, which we discuss later in this chapter.

Designing Around an Existing Patent

Modifying your product so that it is different enough from a competitor's to avoid infringement is called *designing around* the competing product. Patent protection extends only to the scope of the claims contained within the patent; however, the courts often interpret those claims as extending beyond their literal meaning. Courts read patent claims quite broadly in order to prohibit any invention that achieves a substantially equivalent result in substantially the same way. This liberal reading constitutes a legal rule known as the doctrine of equivalents, which basically means that your competitor is not using elements that precisely correspond to your patent claims, but is substituting equivalent elements that perform the same function to achieve the same result.

In a 2002 case called *Festo Corp. v. Shoketsu Kinzoku Kogyo Kabushiki Co., Ltd.*, the U.S. Supreme Court reaffirmed the doctrine of equivalents as providing the necessary certainty and meaningful protection to inventors to encourage their investment in research and development. The Festo corporation brought to market an industrial device whose patent applications had to be amended twice due to insufficient description of two particular claims limitations. Shortly after the amendments had been made, Shoketsu began producing a nearly identical device that did not include the two particular claims limitations that had been disputed in Festo's patent application. Festo, of course, sued them, alleging that Shoketsu's design was an infringement under the doctrine of equivalents. The Court ruled in favor of Festo, explaining that "if patents were always interpreted by their literal terms, their value would be greatly diminished. Unimportant and insubstantial substitutes for certain elements could defeat the patent and its value to inventors could be destroyed by simple acts of copying."

> **It's the scope of the claims in your patent application that will ultimately determine your rights to stop a competitor from marketing a functionally similar product or process.**

Before You Call an Attorney

Patent infringement litigation is one of the more expensive legal endeavors, and not only will you probably need an attorney who specializes in patent law, but you're also likely to need expert witnesses. Depositions and pretrial discovery (the gathering of enough evidentiary support to substantiate a subpoena) can cost hundreds of thousands of dollars. It can cost $1 million, or even more, to litigate a patent dispute. This is why most lawsuits are settled, with letters of negotiation and licensing agreements, well before they reach the courtroom. It is also a reason why it's so important to know how to determine whether someone is infringing your patent, or if you are infringing someone else's.

RESEARCH THE SPECIFICS OF THE COMPETITOR'S PRODUCT Before you even think of calling a lawyer—and starting their billing clock—you need to learn everything you can about the competitor's product. An infringer may have had to learn about your product to copy it, or may have accidentally and independently developed an infringing product. Regardless, you need to know as much as you can about its development to be sure that your case has enough support to justify the massive amount of time and money you are about to invest.

Sometimes it's possible to examine the product itself to get the particulars about its production. However, manuals, advertisements, and information provided by the competitor to its customers can be just as valuable in determining crossover comparisons. You can greatly assist your attorney and keep your legal bill down by carefully preparing a list of each and every similarity you note, as well as any aspects of the product or process that differentiate it from yours. Even if the competitor has added their own features to the product, which has all the elements of your product, the competitor may be infringing by making an unauthorized improvement. But remember, it's the scope of the claims in your patent application that will ultimately determine your rights to stop a competitor from marketing a functionally similar product or process.

ASK YOURSELF IF THE DOCTRINE OF EQUIVALENTS APPLIES When you compare any differences between your product and your competitor's, you need to look carefully at what these differences actually accomplish. For example, if your patent claims an elastic band and your competitor has included a spring, both may accomplish the same purpose and may be considered immaterial modifications under the doctrine of equivalents.

Possible Remedies for Infringement

As Robert Kearns, inventor of the intermittent windshield wiper, demonstrated, individual inventors can and do win lawsuits when their patents are infringed. If you can prove you are a victim of patent infringement, the courts have a fierce array of remedies to make you whole. The availability of these remedies is precisely what gives infringers, large and small, an incentive to settle with you.

Injunctive Relief

Injunctive relief is sometimes viewed as the brass ring in patent litigation. It basically means that the court orders your competitors to stop their infring-ing activity. An injunction may be temporary or permanent. A preliminary injunction is issued on a temporary basis while litigation is pending. The injunction can be made permanent, depending on the outcome of the litigation.

Because injunctions can be financially devastating to a defendant, potentially putting him out of business, courts have very strict standards for their issuance. Generally, a court must determine that all five of the following factors are present before it will issue an injunction:

1. The patent owner must have a substantial likelihood of prevailing on the merits of the case at trial.

2. The patent owner will suffer irreparable harm if the preliminary injunction isn't granted.

3. The patent owner doesn't have another adequate remedy at law that will keep them from suffering the irreparable harm.

4. The threatened harm to the patent owner outweighs the potential harm to the defendant.

5. The granting of a preliminary injunction serves the public's interest.

If a preliminary injunction is granted, there is a strong likelihood that the patent holder will win his or her case and succeed in making the injunction permanent. Thus, many of the issues that would otherwise be resolved at trial may be effectively decided during the hearing for the preliminary injunction.

Raymond Niro: The Ultimate Patent Troll

Raymond Niro is a Chicago-based litigator whose unconventional strategy of hunting for patents (rather than clients) has earned him considerable wealth as well as a dubious title: the original patent troll.

In one of his first big "contingency" cases, Niro and his law partner sued more than forty companies (including technology titans AT&T, IBM, Sony, and Dell) for infringing patents on early answering machine technology that a client had bought from a Japanese inventor. The licensing fees in the litigation earned them $65 million.

Today Niro still runs a patent infringement practice and takes a large percentage of the settlement he earns for his clients on a contingent fee basis. Many of his clients—individual inventors—could never finance patent infringement litigation against big companies on their own.

Monetary Damages

Damages compensate the plaintiff for the costs and economic losses it has suffered as a result of infringement. If the court finds that the infringement was willful, the judge can triple the amount of damages.

When Kodak was found liable for infringing Polaroid's patent with its own instant imaging camera, triple the cost of damages was requested. Polaroid expressed disappointment that they received only a $925 million damage award, which was less than they had hoped for.

Attorneys' Fees

You probably shouldn't initiate a lawsuit with the expectation that you'll be reimbursed for your attorneys' fees. It's the rare exception, rather than the rule, that courts use their discretion to award attorneys' fees to the prevailing party, even in cases of willful infringement.

Nevertheless, patent owners routinely ask for attorneys' fees to be covered in the settlement. Occasionally, a court is so moved by the plight of a patent holder and the egregious nature of a defendant's conduct that such a request is successful.

Important Considerations Prior to Filing Suit

All litigation has risks to the plaintiff and the defendant. You should be prepared to pay the costs of bringing a lawsuit without any certainty that these costs can be recouped. Some law firms will represent a client on a contingent fee, meaning that the fees owed to the attorney are payable only if the suit is decided in your favor, but these firms are very selective about the cases they take on. Also, litigation often breeds more litigation, in the form of civil counterclaims and actions taken on behalf of the Patent Office in examining the legitimacy of your original patent.

What Can You Recover from the Defendant If You Win?

Even the most favorable judgment stated in the most certain terms is no guarantee that you will be paid a dime. Many defendants, faced with the devastating impact of combined injunctive relief and damage awards, simply elect to file bankruptcy. Bankruptcy affords filers (in this case, your defendant) protection from all their creditors, and there may be quite a few of them besides you.

Whom Can You Name as a Defendant in Your Lawsuit?

Federal law permits inventors to cast a wide net when it comes to naming defendants in an infringement suit. It's a good idea to take advantage of this reach, naming as many culpable parties as possible in order to increase your chances of having an action that names responsible, financially solvent defendants who can ultimately foot the bill for any damage awards.

Examples of parties you may sue include the following:

- Retailers who sell the infringing product

- Factories that produced the product

- Purchasers and end users of the product

- Advertisers who encourage the public to infringe your patent or tell them how to do it

Infringement occurs when one of your three exclusive rights as a patent holder is infringed. These three rights include using, making, and selling your patented device or process. There are also three types of infringers: direct, indirect, and contributory. The distinction is important, because ignorance and good faith constitute defenses only if the infringement is contributory.

Direct infringement occurs whenever someone makes, uses, or sells your patented product. Direct infringement can be committed innocently if the infringer has no personal knowledge of your patent. In fact, this is often the case. However, the law provides that a patent owner must mark his or her own product with notice of the patent, or give actual notice, in order to receive monetary damages. (As an inventor, your product and all marketing materials should clearly show that the product is "patent pending" or patented and include the relevant patent numbers.)

Indirect infringement occurs when someone doesn't make, sell, or use your product themselves but instead induces someone else to do so. For example, if someone sells information telling the public how to manufacture your patented product, or how to replicate your patented process, they are liable for indirect infringement.

Contributory infringement occurs when an infringer knowingly sells a component that contributes to the infringement process with the intent to profit from the infringement. The component itself may be in the public domain. For example, you may be liable for contributory infringement if you knowingly supply tennis shoes to a company that is equipping the shoes with lights, violating someone's patent for a light-up gym shoe. The shoe does not in and of itself infringe any patent; however, contributory infringement statutes can actually bring unpatentable matter, like a tennis shoe, within the scope of the Patent Act.

If you believe a competitor has infringed your issued patent, resist the temptation to write a threatening letter.

The law specifies the following three-part test for contributory infringement:

- **Sale:** There must be a sale of the infringing goods.

- **Material component:** The sale must involve a material component of the invention as defined in the claims of the patent.

- **Knowledge:** The defendant must have actual knowledge that he or she is trafficking in infringing goods.

For example, in the early 1960s Ford Motor Company manufactured hundreds of thousands of convertibles with a popular retractable top. The problem was that Ford had neglected to enter into a valid license with the convertible tops' patent holder. Each manufacturer of the convertible tops would have been liable for contributory infringement if it could have been proven they had actual knowledge of the infringement. Since the statute requires "knowing" infringement, however, the top manufacturers were held liable only for continued manufacture of those convertible tops after having received notification of the invalid license agreement with Ford.

Also, in the late 1980s, there was a huge black market in the United States for counterfeit General Motors truck parts. Retailers who knowingly resold the infringing parts could have been named as defendants in an infringement suit had such a suit been filed.

THE STATUTE OF LIMITATIONS AND THE NEED TO ACT PROMPTLY In cases of patent infringement, there is no *statute of limitations,* which is the amount of time one has to file a suit after a crime has been committed. However, only damages from the six years immediately preceding the filing date can be recovered. So, if you find that your patent was infringed twenty years ago, for example, you may still sue, but whatever damages you are awarded will cover only the six years leading up to the time of your suit. This does not mean you should sit back and watch someone infringe on your patent and wait until they sell millions of units before taking action. Because if you fail to contact the infringer, the court may decide that you allowed the continued infringement and did not act reasonably to minimize your damages. Since damages are assessed at the court's discretion, this may affect the amount of your award.

WHERE CAN YOU FILE YOUR LAWSUIT?

Generally, if an action is for patent infringement, you can bring your lawsuit "in the judicial district where the defendant resides, or where the defendant has committed the act of infringement and has a regular and established place of business." If you do business in New Mexico, for instance, you don't want to travel to Alaska to defend yourself in a lawsuit that's filed under Alaska law. Location can dramatically affect the cost and ability to defend, and therefore the ultimate outcome of a lawsuit.

Jurisdiction is the legal authority of a particular court to hear a case. Federal courts have jurisdiction over lawsuits involving patents. This is because the authority to regulate patents arises under the U.S. Constitution.

The federal courts, however, have exclusive jurisdiction only over disputes that specifically involve patents' enforcement, infringement, and validity. If there is a contractual dispute involving a patent license agreement, a state court has the right to hear the matter, since it is state law that governs the contract.

Your Burden of Proof

In any lawsuit, the person who files the suit has a *burden of proof*, which is, essentially, evidence that you must legally establish in order to win your lawsuit. If a defendant brings counterclaims, or alleges that your patent is invalid, the defendant has the burden of proving those things.

There are different evidentiary standards that plaintiffs and defendants must each meet, depending on the type of action being filed. As the plaintiff in a patent infringement suit, you must show that the *preponderance of evidence* points to the conclusion that the subject patent has been infringed; in other words, that there is more evidence that the patent has been infringed than that it has not been.

If a defendant alleges as a defense that your patent is invalid or unenforceable, they have a higher burden of *clear and convincing proof* than if they bring only counterclaims against you. The clear and convincing proof standard means that the conclusion drawn from the evidence is highly probable and free from reasonable doubt.

Resolving Infringement Issues Without Litigation

If you believe a competitor has infringed your issued patent, resist the temptation to write them a threatening letter. Doing so can subject *you* to a lawsuit, filed in the jurisdiction in which they are located, to have your patent declared invalid. Instead, you should obtain a sample of their product and take it to your patent attorney. If the attorney determines that the competitor is infringing on at least one claim of your patent, he or she will craft an appropriate communication. Often the infringement is inadvertent, and the competitor may be happy to pay you a royalty and even obtain the product from you in the future if they currently get it from another source. This is another good reason to keep communication amicable at the beginning.

A "cease and desist" letter to the infringer should do the following:

■ Ask the infringer to stop their unauthorized manufacture, use, or sale of your product, and to pay you royalties for past infringement.

A patent is unenforceable and invalid if the device or process was sold or used a year prior to the patent's being filed.

■ State the amount or range of royalties you expect to be paid, or the way in which your royalties should be computed.

■ Offer the infringer the opportunity to enter into a valid license agreement with you in the future.

Defending Yourself If You Are Accused of Infringement

It's definitely a shock when your morning mail brings a letter notifying you that your invention falls within someone else's patent claims. Usually these letters are carefully worded as an "invitation" to take a license.

If you believe your product is clearly distinct from and dissimilar to any other, your first impulse may be to take a wait-and-see approach. As discussed earlier, the plaintiff in an infringement action must meet a heavy burden of proof (supply a good deal of evidence) in order to move forward with the case. However, it's in your interest to quickly assess the potential impact on your business, and to consider which of the following positions you're going to take in response to that letter.

Position #1: "I am not infringing the patent."

After taking a good look at your product and your competitor's, you may decide that they are not the same at all. In that case, it's important that you distinguish your product from what's covered in your competitor's patent. Remember, in order to prevail in a lawsuit, the patent holder must show that your invention includes each and every element of at least one claim in their patent. You can get a copy of their patent off the Patent and

Trademark Office website: www.uspto .gov. The patent date and number may have been included in the initial cease and desist letter; if not, you can write back and request it from the patent owner. Google. com also has a Patent Search application that features the entire patent collection from the USPTO from the 1790s to the near present.

Position #2: "The patent is invalid."

The owner of a patent places himself at risk by sending you the cease and desist letter: They have opened themselves to your challenging the validity of their patent.

You can challenge the enforceability of a patent in the following ways:

Initiate an official reexamination. You can ask the PTO to reexamine your competitor's patent in view of prior art that was not considered at the time of the initial application and which you've now identified. The cost of the process is fairly substantial, but it's usually cheaper than defending against an infringement suit.

Provide evidence that the product or process was sold or used one year prior to the filing of the patent. A patent is unenforceable and invalid if the device or process was sold or used a year prior to the patent's being filed. Go back into your research files to see if you can find any of your competitor's advertising, customer contracts, records

of manufacturing costs, or other documentation that would provide evidence of sale or use one year prior to the filing of their patent.

Demonstrate that the patent holder committed a fraud during the patent process. It is considered a fraud sufficient to invalidate a patent application if the patent owner failed to disclose relevant facts to the patent examiner during the patent process. Relevant facts may include prior art, information about previous patents, or the best mode (that the inventor disclosed what he thought was the best method of producing his invention). You may be able to uncover these matters by reviewing the patent file with your attorney.

Show the PTO that the invention disclosed doesn't work or that the explanation of the invention is incomplete. The patent application must describe a working implementation of the invention, and the application must also sufficiently describe to someone "skilled in the art" how they could make the invention. If you can demonstrate that the patent owner has failed to do either of these things, the Patent Office may declare the patent invalid.

Provide evidence of misuse of the patent. If the patent holder has misused the patent, the PTO will invalidate the patent retroactively. A finding of misuse is rare and has most frequently involved antitrust violations. However, showing that the patent owner has used his invention for any illegal purpose may, theoretically, be used to invalidate the patent.

Position #3: "The patent is unenforceable."

A perfectly good patent can be rendered unenforceable for a variety of legal faux pas.

Here are a few of the ways a patent holder can shoot himself in the foot:

False marking: False marking is the incorrect marking of patented products with numbers that don't cover the patent that was granted. Patent holders are required to mark their product materials to put the public on notice of the patent. Failure to do so can negate their right to damages. Including an incorrect patent number is considered false marking and can render the patent unenforceable.

Unreasonable delay in bringing a lawsuit: The law frowns upon a patent owner who ignores infringing activity. Suppose a patent holder watches a manufacturer develop, market, and sell a product for a decade, and then decides to sue just when the product line catches on and begins turning a profit. The defendant can argue that the patent holder has been aware of the infringing activity for years, has taken no action until this time, and therefore should be stopped from enforcing the patent.

Patenting an invention that doesn't actually work: If a patent holder's device or process doesn't work, the patent may be unenforceable against a competitor who develops a working implementation.

Stopping Foreign Knock-Offs: The Customs Service

As mentioned earlier, a patent prevents others from making, using, or selling your invention. It also protects you from others importing an infringing item into the United States. The U.S. Customs Service and the International Trade Commission have an official obligation to look out for the interests of U.S. businesses that hold valid patents and trademarks. Their services aren't necessarily alternatives to suing in federal court, but they can be useful and appropriate under the right circumstances.

You can request that the Customs Service conduct an import survey for the duration of two, four, or six months (ranging in cost from $1,000 to $2,000). The Customs Service will provide you with the address of any importer whose goods appear to infringe on your patent. Sometimes this procedure has the collateral effect of delaying the importation of the infringing goods. To obtain an application for an import survey, you can write to the Commissioner of Customs, Attention IPR Branch, Room

2104, U.S. Customs Service, 1301 Constitution Avenue, Washington, DC 20229.

If your patent is particularly valuable and you need to take steps to stop its infringement on a global scale, you can bring a proceeding before the International Trade Commission. This type of proceeding can be expensive, but it is very powerful. It enables you to have the importation of the product stopped and quarantined at every point of entry, including all American seaports as well as borders with Canada and Mexico.

You have now digested a tremendous amount of legal advice and information that would scare even the most seasoned inventor. The truth is, most patents are not infringed, and when they are, reasonable actions can be taken to remedy the situation. Understanding the potential risks of inventing is critical to assessing the value of the potential rewards. Patent litigation is one of those risks, regardless of how remote.

The Constitution provided for patent protection as an incentive for innovation, to develop new and useful ways to make our lives better and advance society. The U.S. patent system is the best in the world, but certainly it is far from perfect. The better you understand your rights and the ways the system can be used to your benefit, the better

the inventor you become. Having all the information in advance allows you to achieve this.

As we look back on history, we marvel at the inventions that were developed by individuals just like you—risk takers who pursued their ideas and, using the patent system, benefited from the protection afforded by the United States Constitution. One day, hopefully, future generations will look at one of your inventions as they ponder its contribution to society.

It's hard to imagine what our lives would be like without innovation. All of contemporary culture has benefited from the invention process. The lights that illuminate our homes and cities, the vehicles that move us from place to place, the phones and computers that keep us connected to family and friends, and the countless other inventions that we give hardly any thought to, but that make modern living easier.

What all these innovations have in common is that they all began as an idea by an individual (not a huge corporation)—someone who said, "There has to be a better way," who followed through with that idea until they were able, through hard work, sacrifice, and investment, to turn that idea into a marketable and sellable product. These innovations changed our lives and made fortunes for many of their creators.

Becoming the Next Edison

Your Personal Road Map

The most important takeaway from this book is the understanding that inventing is a process of problem solving. It is a process of transforming a raw idea into a finished project in a systematic way. Your own path will depend on your field and your personal vision of accomplishment. The focus of this book has been consumer products and the concepts that make our lives dynamic and colorful. But many of the discussions in these pages apply to inventing and refining concepts that go beyond consumer markets. The principles apply equally to scientific advancements, complex technologies, and ideas that shape the future of the world. All inventors have the gift of creative vision, as well as the need for resources that will help take their inventions to the next level.

Chart Your Course

hether you are dreaming of creating a plastic toy or an alternative energy system, there are important milestones you'll encounter on the road from raw idea to refined invention. Below is a map of sorts, to give you a collected overview of some of the things you can expect and what questions you should ask yourself along the way. It's important to get to "yes" before advancing to the next phase.

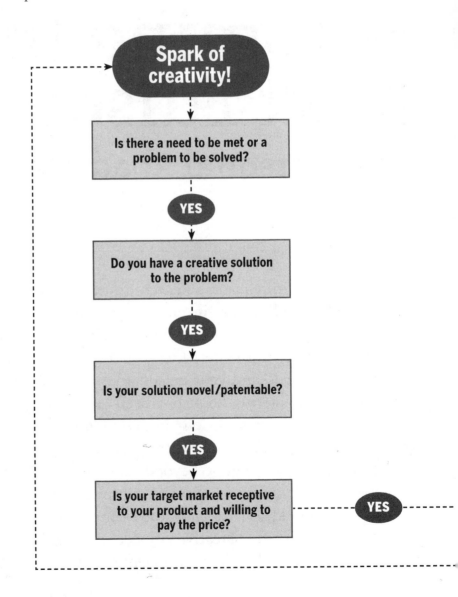

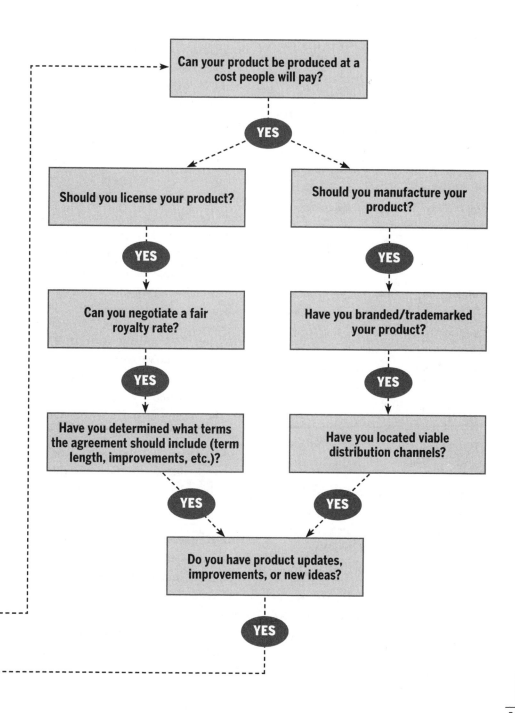

Lessons Learned from This Book

As you sift through your ideas to determine the ones that can be successful, your essential tasks will be:

❏ **Realizing that inventing is a (creative) process.** Inventing is a creative calling. However, unbridled creativity is a far cry from a structured plan for bringing an actual product to market. It takes careful calculation of risks, consideration of costs and benefits, and assessment of your target market.

❏ **Identifying your target market/ user base.** Who is your customer? You need to be able to accurately define this person and then be able to quantify the number of these customers in the market. Your customer is not everyone, and thinking as much is foolish. If there are different groups who make up your customer base, start out working with the largest.

❏ **Figuring out how to fill the need.** Once you've found the target market, design a product that fills your newly identified niche and solves a problem at a price people are willing to pay.

❏ **Protecting your solution.** After you've done the research, make sure you have something you can protect with a patent before you embark on developing a solution. Moving forward with your invention is a financial investment and it should be protected.

❏ **Funding your dream.** Bringing a new product to market is rewarding both financially and intellectually. But even the most brilliant concept is a risk. Will the market be ready? Can you afford to invest what it takes to fund the patent and prototyping processes? And even if you can, should you? It can be tricky to navigate the ins and outs of bootstrapping and other funding options (it's unlikely that an angel investor is going to swoop down to finance new ideas brought to them by strangers), but have the confidence that you can approach the people who rightly believe in your ideas and stand to benefit from your ingenuity.

❏ **Pitching and selling your invention.** Once you've turned your raw idea into a marketable product, be prepared to make your pitch for it. As the inventor, you are the most knowledgeable about the product, the most passionate, and will benefit the most financially. Get out there and be the evangelist for your invention.

❏ **Profiting from your idea in a global marketplace.** Patents do not guarantee profits, but if you have followed the system we've outlined in the book, you will be prepared to launch and protect your invention in a global marketplace and will have learned the process of inventing for successive inventions. You are well on your way to being the next Edison.

❑ **If the numbers don't add up . . . don't do it.** So many inventors fall in love with their idea and, as a result, follow the path blind to the potential consequences. Successful inventing requires discipline. It takes discipline to follow through and execute, as well as to know when not to. There is no statute of limitations on being an inventor. If the rewards of the first idea don't justify the risks, either wait until they do or just keep looking for another. Great inventors rarely have only one idea; they have many ideas and act on the really profitable ones. That is what makes them great inventors.

Wishing You Luck—
Let Us Know How You're Doing

The process of inventing is evolutionary, and the information that you will need to stay in front of your competition is in constant flux. Do your best to stay current on trends and technology, but most importantly, never stop learning. Surround yourself with people who have done it before or who compliment the skills you have. With the launch of this book, we have also launched companion websites, www.independentinventorshandbook.com and www.freepatentlaw.com, to provide new and useful information to you as it develops.

We'd love to hear from you! Good luck.

Authors Jill Gilbert Welytok and Louis J. Foreman

Appendix A

Confidential Nondisclosure Agreement (NDA)

There is no set formula for an NDA. You should always consult an attorney with respect to your specific invention. This is a sample form only and does not protect you from the one-year time bar for getting a patent after making an offer for sale (see the "on-sale bar," page 232).

CONFIDENTIAL NONDISCLOSURE AGREEMENT

Between: [Company name and address] "Receiving Party"

and [Your name and address] "Disclosing Party"

1. On the understanding that both parties are interested in meeting to consider possible collaboration in developments arising from [your name]'s ("Disclosing Party") [describe your invention here], it is agreed that all information whether oral, written or otherwise, that is supplied to the Receiving Party in the course of any meeting shall be treated as confidential by the Receiving Party.

2. The Receiving Party undertakes not to use the information for any purpose, other than for the purpose of considering collaboration, without obtaining the written agreement of the Disclosing Party.

3. This Agreement applies to both technical and commercial information communicated by either party.

4. This Agreement does not apply to any information in the public domain or which the Receiving Party can show was either already lawfully in their possession prior to its disclosure by the Disclosing Party or acquired without the involvement, either directly or indirectly, of the Disclosing Party.

5. Neither party to this Agreement shall retain any documents or items connected with the disclosure after collaboration has ceased.

6. No disclosure made by the Receiving Party shall create any license, title, or interest in respect of any Intellectual Property Rights of the Disclosing Party.

7. After ten years from the date hereof, each party shall be relieved of all obligations under this Agreement.

Signed _____ "Disclosing Party"   Date _____
              [Name]

Signed _____ "Receiving Party"   Date _____
              [Name]

Appendix B

Performing a Basic Patent Search

There are plenty of good reasons to do a patent search. In order to be patentable, your invention must be useful (which most inventions are), novel, and nonobvious. To determine whether your patent meets these criteria, you need to do a basic patent search to see what is already out there, and assess how different or similar your idea may be. In light of the existing patents, will the examiner conclude that your innovation is "obvious"? This is a very subjective assessment, and examiners may differ.

The first step in the search process is to come up with key words that will likely appear in any issued patent or published application related to your invention. (Patent applications are published 18 months after they are filed, regardless of whether they are abandoned or if the patent has not yet been issued because the examination is still being completed.)

Two of the best places to start your patent search are Google Patent Search and the United States Patent and Trademark Office. (Both are free resources.)

Google Patents is located at www.google.com/intl/en/options, and has a searching interface that will be familiar to most Google users. You can enter keywords for search terms, or the name of an inventor.

The advantage of Google Patents, in addition to its being an incredibly efficient search engine, is that you can download and print out a complete copy of the patent and applications for free. The U.S. Patent and Trademark Office charges a fee for obtaining complete electronic copies of patents.

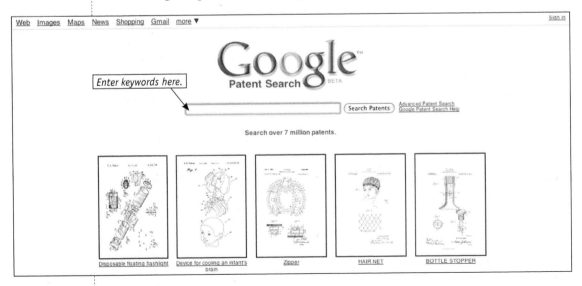

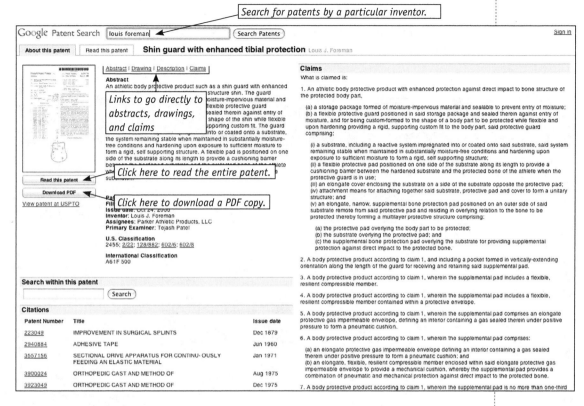

Just keep in mind that pending patent applications are not available in the search results for Google Patents, nor are patents issued within the few months prior to the search. You can, however, find these documents on the USPTO website located at www.uspto.gov, shown on the next page.

To search the USPTO site, follow these steps.

1. Point your browser to www.uspto.gov.

2. Click on the "Patents" link on the left-hand side of the screen.

3. Click on the third search option called "Search Patents."

4. If you want to search published, or "issued" patents, click on one of the search links in the green box on the left-hand side of the screen. To search pending (published) applications, click on one of the search links in the yellow box on the right-hand side of the screen.

Remember that a majority of published applications are not required to be published until 18 months after filing. This can be a long time in fast-moving markets like consumer products. Therefore, we always recommend that if you think someone else has thought of or is marketing your invention, do a basic Google search. Sometimes companies publish their Web-based advertising and actually launch products much sooner than their patent applications are published.

Patents.

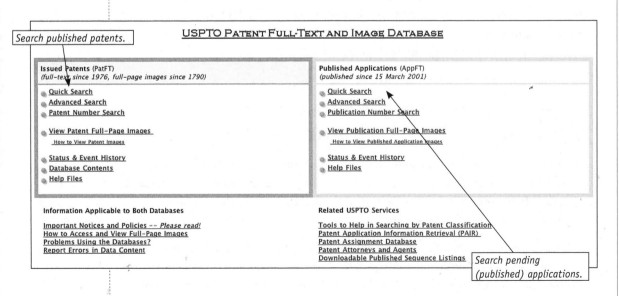

Search published patents.

Search pending (published) applications.

Appendix C: Excerpts from a Sample Patent

US007389897B2

In the U.S. the inventor's name must be on the application, although an assignee of the inventor's rights (such as the company for which he or she works) may also be listed.

(12) **United States Patent**
Pistiolis et al.

(10) Patent No.: **US 7,389,897 B2**
(45) **Date of Patent:** Jun. 24, 2008

Patents issued after 1995 expire 20 years after the filing date. Patents issued prior to 1995 expire the later of 20 years from the filing date or 17 years after the issue date.

(54) **BABY BAG CONVERTIBLE INTO BABY CARRIER**

(75) Inventors: **Maria P. Pistiolis**, Charlotte, NC (US); **Louis J. Foreman**, Huntersville, NC (US); **Daniel L. Bizzell**, Charlotte, NC (US); **Ian D. Kovacevich**, Charlotte, NC (US)

(73) Assignee: **Snuggle Nest, LLC**, Charlotte, NC (US)

(*) Notice: Subject to any disclaimer, the term of this patent is extended or adjusted under 35 U.S.C. 154(b) by 188 days.

(21) Appl. No.: **11/160,001**

(22) Filed: **Jun. 3, 2005**

(65) **Prior Publication Data**
US 2005/0210594 A1   Sep. 29, 2005

Related U.S. Application Data

(63) Continuation-in-part of application No. 11/051,970, filed on Feb. 3, 2005, now abandoned.

(60) Provisional application No. 60/541,546, filed on Feb. 3, 2004.

(51) **Int. Cl.**
A47D 13/02   (2006.01)
(52) **U.S. Cl.** 224/158; 224/576; 224/577
(58) **Field of Classification Search** 224/158, 224/576, 577, 159; 190/107, 127; 150/130; A47D 13/02
See application file for complete search history.

Prior patents the examiner has read and found distinguishable from this patent must be listed.

(56) **References Cited**
U.S. PATENT DOCUMENTS
3,096,917 A * 7/1963 Gudiksen 294/140

3,297,119 A *	1/1967	Viol	190/8
4,566,130 A	1/1986	Coates	
D288,613 S *	3/1987	James et al.	D25/56
4,712,258 A *	12/1987	Eves	5/424
4,761,032 A *	8/1988	Sanchez et al.	297/229

(Continued)

FOREIGN PATENT DOCUMENTS
JP   04067815 A * 3/1992

(Continued)

OTHER PUBLICATIONS

Exhibits generated by Applicant comprising Exhibit 1 through Exhibit 23 each representing a photograph of a baby bag convertible into a baby carrier as used in public in Aug. of 2003.

(Continued)

Primary Examiner—Nathan J. Newhouse
Assistant Examiner—Jack H Morgan, Jr.
(74) *Attorney, Agent, or Firm*—Tillman Wright, PLLC; Chad D. Tillman; James D. Wright

An abstract is a summary of the claims, but generally has no legal effect on the interpretation of the patent or application.

(57)   **ABSTRACT**

A baby bag that is convertible to a baby carrier includes a base that is movable between a folded configuration for storage and transport of the baby bag, and an unfolded configuration for receiving a baby. In an embodiment, a harness is connected to the base for restraining a baby received upon the base such that a back of the baby is maintained in close proximity to the base. In the same or another embodiment, a shoulder strap is attached to the base for suspension of the base, whereby a baby received upon the base may be carried by a person utilizing the shoulder strap.

20 Claims, 26 Drawing Sheets

Drawings help you determine the similarity of unrelated products and inventions at a glance and legally must show each and every element of the completed invention.

US 7,389,897 B2

Page 2

U.S. PATENT DOCUMENTS

4,999,863	A	*	3/1991	Kane 5/98.1
5,439,154	A		8/1995	Delligatti
5,551,108	A	*	9/1996	Butler, III 5/655
5,826,714	A		10/1998	Martin
6,168,022	B1		1/2001	Ward et al.
6,213,304	B1		4/2001	Juliussen
6,298,993	B1		10/2001	Kalozdi
6,390,260	B1		5/2002	Roegner
6,954,955	B2	*	10/2005	Brewin et al. 5/655
2001/0040177	A1	*	11/2001	Hammond 224/188
2002/0011503	A1	*	1/2002	Hwang 224/160
2002/0050503	A1		5/2002	Spero
2005/0051582	A1	*	3/2005	Frost 224/160
2006/0096031	A1	*	5/2006	Foster 5/655

FOREIGN PATENT DOCUMENTS

JP		2000296035	A	* 10/2000

WO   WO02098263 A1 * 12/2002

OTHER PUBLICATIONS

Great Expectation, "Great Expectations Baby Traveller Range of Diaper / Nappy Bags", http://www.greatexpectations.com.au/interiorpages/babytraveller_range.shtml, publication date unknown <accessed May 10, 2005>.

Samsonite, "Samsonite Juvenile Products", http://www.samsonitejuvenile.com/products2.php?collectionID= 1, publication date unknown <accessed May 10, 2005>.

Small Fry Design, "Small Fry Design—Infant and Toddler Toys and Accessories", http://www.smallfrydesign.com/, publication date unknown <accessed May 10, 2005>.

Small Fry Design, "Small Fry Design—Infant and Toddler Toys and Accessories: Travel N Trundle, Mobile Clip, My Playh . . .", http://www.smallfrydesign.com/accessories.htm, publication date unknown <accessed May 10, 2005>.

Unique Baby Products, "Buggy Bagg—BuggyBagg shopping cart liner", http://www.uniquebabyproducts.com/buggybagg.htm, publication date unknown <accessed May 10, 2005>.

* cited by examiner

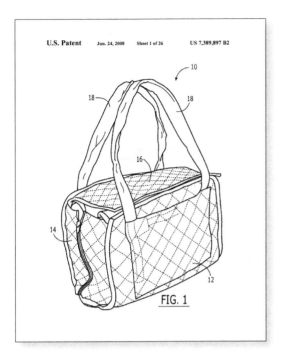

FIG. 1

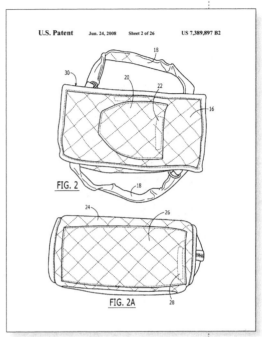

FIG. 2

FIG. 2A

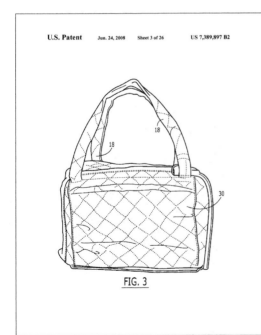

FIG. 3

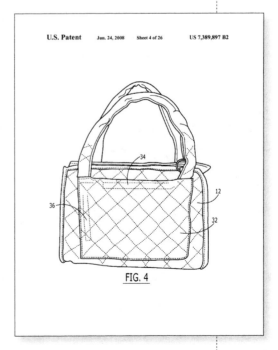

FIG. 4

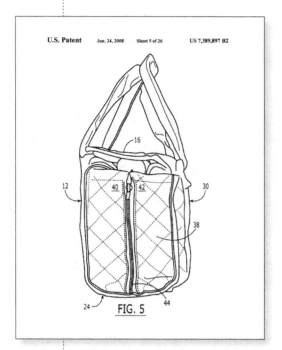

FIG. 5

FIG. 6

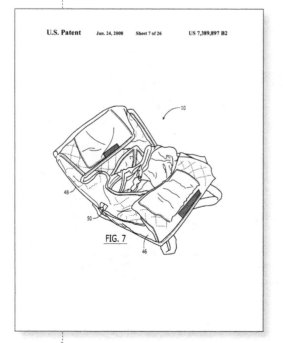

FIG. 7

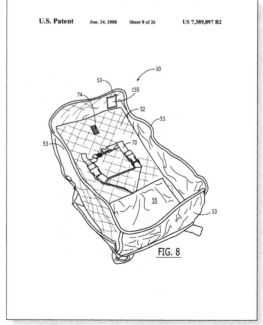

FIG. 8

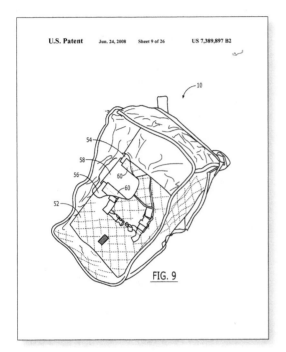

FIG. 9

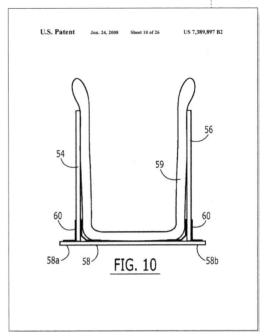

FIG. 10

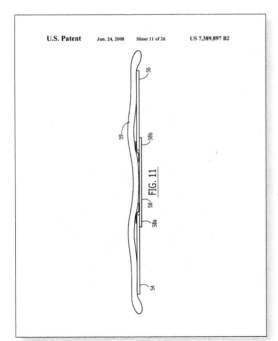

FIG. 11

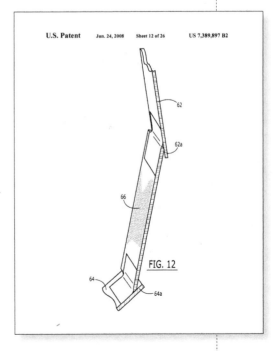

FIG. 12

FIG. 13

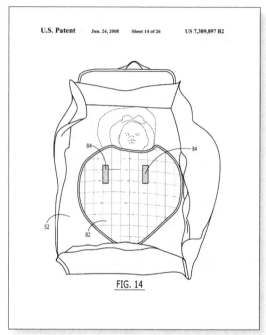

FIG. 14

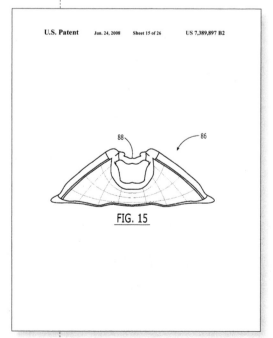

FIG. 15

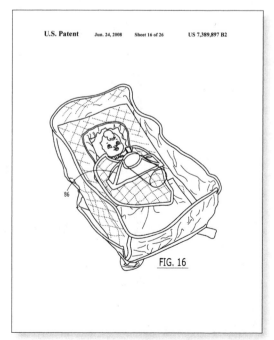

FIG. 16

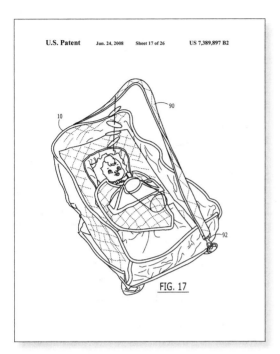

FIG. 17

FIG. 18

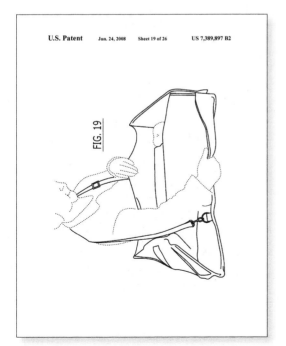

FIG. 19

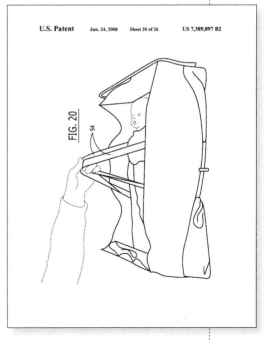

FIG. 20

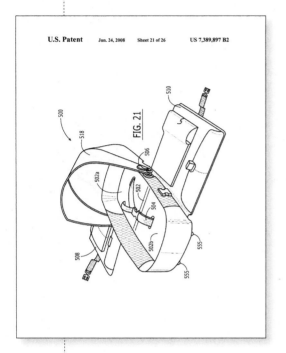

U.S. Patent   Jun. 24, 2008   Sheet 21 of 26   US 7,389,897 B2

FIG. 21

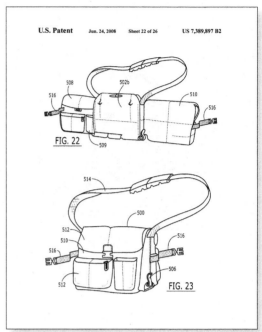

U.S. Patent   Jun. 24, 2008   Sheet 22 of 26   US 7,389,897 B2

FIG. 22

FIG. 23

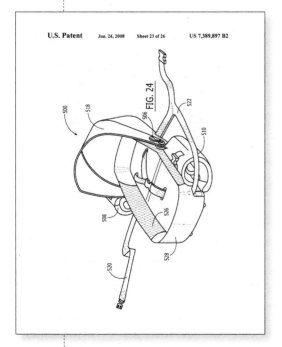

U.S. Patent   Jun. 24, 2008   Sheet 23 of 26   US 7,389,897 B2

FIG. 24

U.S. Patent   Jun. 24, 2008   Sheet 24 of 26   US 7,389,897 B2

FIG. 25

U.S. Patent Jun. 24, 2008 Sheet 25 of 26 US 7,389,897 B2
FIG. 26
FIG. 27
FIG. 28
530
532
534
506
556
557

U.S. Patent Jun. 24, 2008 Sheet 26 of 26 US 7,389,897 B2
FIG. 29
FIG. 30
630
632
634
636
638
640
642
644
646
558
559

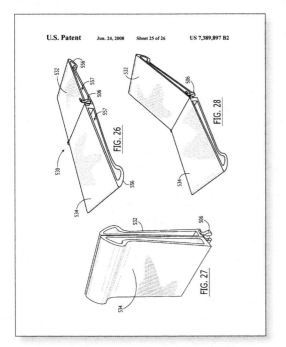

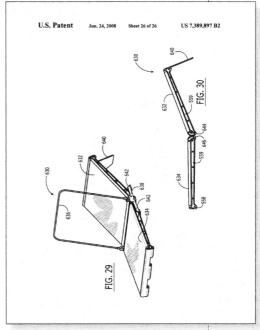

The claims part of the application is the legal description of the invention, much like a real estate deed. In order to infringe the patent, another product must have all of the elements of at least one claim of the patent.

US 7,389,897 B2

1

BABY BAG CONVERTIBLE INTO BABY CARRIER

CROSS-REFERENCE TO RELATED APPLICATIONS

This application is a continuation-in-part of and claims priority to U.S. patent application Ser. No. 11/051,970, filed Feb. 3, 2005 now abandoned, which is hereby incorporated herein by reference, and which application is a nonprovisional patent application of and claims priority to U.S. Provisional Patent Application No. 60/541,546, filed Feb. 3, 2004, which is hereby incorporated herein by reference.

COPYRIGHT STATEMENT

All of the material in this patent document is subject to copyright protection under the copyright laws of the United States and of other countries. The copyright owner has no objection to the facsimile reproduction by anyone of the patent document or the patent disclosure, as it appears in the Patent and Trademark Office patent file or records, but otherwise reserves all copyright rights whatsoever.

The background section explains the scientific and technical files and history of other related inventions.

BACKGROUND OF THE INVENTION

A variety of bags intended for use in carrying childcare accessories is commercially available. Available bags are useful for storing and transporting accessories and can be seen in many public environments where care providers such as parents and young children are seen. For example, young mothers carrying children and bags of diaper-changing supplies can often be seen in shopping districts, public parks, and in public transportation facilities. A care provider can face considerable physical prowess and balance challenges in carrying both a child and a typical care accessory bag throughout the course of any typical daily outing. Yet other challenges arise when a diaper needs to be changed. Even if a diaper changing station is available, such as in a public restroom, a parent may have difficulty while trying to comfort and securely hold a child on the hard surface of such a station while fumbling through a bag to find needed supplies. The cleanliness of such a surface and the overall public restroom environment is yet another concern for such a parent and carrying a pad or the like to isolate the child and the supplies only adds further burden in carrying and placing the pad. Indeed, often are seen awkward ad hoc situations in automobiles and lobbies where diaper changing stations are unavailable and a care provider struggles to feed a child or change a diaper without a convenient platform for the execution of the task. Child carriers are available to securely cradle a child but typical carriers are fashioned such that a child essentially must be removed from the carrier for a diaper changing, further complicating the balancing and juggling actions of the care provider.

SUMMARY OF THE INVENTION

The present invention includes many aspects and features. Moreover, while many aspects and features relate to a bag and a carrier, and are described herein in the context of a baby bag that is convertible to a baby carrier, the present invention is not limited to use only in transporting a bag, contents thereof, and a baby. For example, when a bag according to the present invention is converted to a carrier, the carrier is useful as a baby bed and as a diaper changing station without regard to whether a baby is actually carried about or transported, as will

2

become apparent from the following summaries and detailed descriptions of aspects, features, and one or more embodiments of the present invention.

Accordingly, a first aspect of the present invention relates to a baby bag that is convertible to a baby carrier. In one or more examples, a baby bag includes a base and a harness connected to the base. Furthermore, the base is movable between a folded configuration for storage and transport of the baby bag, and an unfolded configuration for receiving a baby, and the harness is configured to restrain a baby received upon the base.

Several variations according to the first aspect relate to the base having a first member hingedly attached to a second member. In one or more examples, moving the base from the folded configuration to the unfolded configuration includes rotating the first and second base members relative to one another. Optionally, overextension of the base is prevented by abutment of a portion of the first base member with the second base member. In another option, an angle of greater than ninety degrees is defined between the first base member and the second base member when the base is in the unfolded configuration.

Another variation relates to a shoulder strap that is removably attached to the base.

Another variation relates to a base member being cushioned for receiving a baby in comfort.

Other variations relate to pockets. In one or more examples, the baby bag includes a plurality of pockets, one of which may be disposed on a member that is connected to the base for hinging movement. Optionally, a pocket is dimensioned to receive a mobile phone. In another option, a pocket dimensioned to receive a baby bottle is disposed on the outside of the baby bag when the base is in the folded configuration. In yet another option, a pocket is dimensioned to receive an umbrella when the base is in the unfolded configuration for shading a baby.

Another variation relates to a canopy for covering a baby when received on the base.

Another variation relates to a collapsible wall attached to and extending around the perimeter of the base for encircling a baby received upon the base when the base is in the unfolded configuration, the wall collapsing toward the base when the base is moved into the folded configuration. In at least one example, the collapsible wall includes sections comprising mesh.

Another variation relates to a cushion for partially surrounding a baby's head when received upon the base. In at least one example, the cushion is removably attached to the base via hook-and-loop fasteners.

Another variation relates to a blanket for covering a baby received upon the base. In at least one example, the blanket is removably attached to the base via hook-and-loop fasteners.

Another variation relates to an anchor strap connected to the base for receipt therethrough of a belt.

A second aspect of the present invention relates to a method of using the convertible baby bag of the first aspect for carrying a baby. This method includes converting the baby bag to the baby carrier in the unfolded configuration, placing a baby upon the base, restraining the baby to the base using the harness, and carrying the base with the baby restrained to the base.

A third aspect of the present invention relates to a baby bag that is convertible to a baby carrier. In one or more examples, a baby bag includes a base and a shoulder strap attached to the base. Furthermore, the base is movable between a folded configuration for storage and transport of the baby bag, and an

US 7,389,897 B2

3

unfolded configuration for receiving a baby. Moreover, a baby received upon the base may be carried by a person utilizing the shoulder strap.

A fourth aspect of the present invention relates to a method of using the convertible baby bag of the third aspect for carrying a baby. This method includes converting the baby bag to the baby carrier by moving the base into the unfolded configuration, placing the baby upon the base, and suspending the base and the baby by the shoulder strap.

BRIEF DESCRIPTION OF THE DRAWINGS

These and other aspects and features of the invention will be more readily understood upon consideration of the attached drawings and of the following detailed description of particular embodiments of the present invention.

FIG. **1** is a perspective view of a convertible baby bag according to a first embodiment of the present invention.

FIG. **2** is a top view of the baby bag of FIG. **1**.

FIG. **2A** is a bottom view of the baby bag of FIG. **1**.

FIG. **3** is a side view of the baby bag of FIG. **1**.

FIG. **4** is an opposite side view of the baby bag of FIG. **1**.

FIG. **5** is an end view of the baby bag of FIG. **1**.

FIG. **6** is an opposite end view of the baby bag of FIG. **1**.

FIG. **7** is a perspective view of the baby bag of FIG. **1** converting to a baby carrier.

FIG. **8** is a perspective view of the baby bag of FIG. **1** converted to a baby carrier.

FIG. **9** is a perspective view of the baby carrier of FIG. **8** showing an arrangement of rigid members thereof.

FIG. **10** is an end view of the rigid members of FIG. **9**, wherein the rigid members are shown in a folded configuration.

FIG. **11** is an end view of the rigid members of FIG. **9** shown in an unfolded configuration, wherein overextension beyond the unfolded configuration is prevented.

FIG. **12** is a partial perspective view of another arrangement of rigid members for a convertible baby bag, wherein the rigid members are shown in a partially folded configuration.

FIG. **13** is a perspective view of the baby carrier of FIG. **8** with a baby received thereon.

FIG. **14** is another perspective view of the baby carrier of FIG. **13**.

FIG. **15** is a side elevational view of a bottle support according to an embodiment of the invention.

FIG. **16** is a perspective view of the baby carrier and baby of FIG. **13**, wherein a baby bottle is supported by the bottle support of FIG. **15**.

FIG. **17** is a perspective view of the baby carrier and baby of FIG. **13**, wherein the carrier is suspended from a shoulder strap.

FIG. **18** is a perspective view of the baby carrier and baby of FIG. **13** carried by a person utilizing the shoulder strap as in the arrangement of FIG. **17**.

FIG. **19** is a perspective view of the baby carrier of FIG. **13** carried by a person utilizing the shoulder strap in an alternative arrangement.

FIG. **20** is a perspective view of the baby carrier of FIG. **13** supported by other straps.

FIG. **21** is a perspective view of a second embodiment of the present invention, wherein a convertible baby bag is shown in an unfolded configuration to define a baby carrier.

FIG. **22** is a perspective view of the convertible baby bag of FIG. **21**, wherein the base thereof is shown in a folded configuration.

4

FIG. **23** is a perspective view of the convertible baby bag of FIG. **21** in a fully folded configuration.

FIG. **24** is another perspective view of the convertible baby bag of FIG. **21** in an unfolded configuration defining a baby carrier.

FIG. **25** is a perspective view of the baby carrier of FIG. **24** being carried by a person utilizing a shoulder strap assembly.

FIG. **26** is a perspective view of a hinging frame of a convertible baby bag, wherein the hinging frame is shown in an unfolded configuration.

FIG. **27** is a perspective view of the hinging frame of FIG. **26**, wherein the hinging frame is shown in a folded configuration.

FIG. **28** is a perspective view of the hinging frame of FIG. **26**, wherein the hinging frame is shown in an inclined configuration.

FIG. **29** is a perspective view of another embodiment of a hinging frame of a convertible baby bag, wherein the hinging frame is shown in an inclined configuration.

FIG. **30** is another perspective view of certain components of the hinging frame of FIG. **29**.

DETAILED DESCRIPTION

As a preliminary matter, it will readily be understood by one having ordinary skill in the relevant art ("Ordinary Artisan") that the present invention has broad utility and application. Other embodiments also may be discussed and described as being in additional illustrative purposes in providing a full and enabling disclosure of the present invention. Moreover, many embodiments, such as adaptations, variations, modifications, and equivalent arrangements, will be implicitly disclosed by the embodiments described herein and fall within the scope of the present invention.

Accordingly, while the present invention is described herein in detail in relation to one or more embodiments, it is to be understood that this disclosure is illustrative and exemplary of the present invention, and is made merely for the purposes of providing a full and enabling disclosure of the present invention. The detailed disclosure herein of one or more embodiments is not intended, nor is to be construed, to limit the scope of patent protection afforded the present invention, which scope is to be defined by the claims and the equivalents thereof. It is not intended that the scope of patent protection afforded the present invention be defined by reading into any claim a limitation found herein that does not explicitly appear in the claim itself.

Thus, for example, any sequence(s) and/or temporal order of steps of various processes or methods that are described herein are illustrative and not restrictive. Accordingly, it should be understood that, although steps of various processes or methods may be shown and described as being in a sequence or temporal order, the steps of any such processes or methods are not limited to being carried out in any particular sequence or order, absent an indication otherwise. Indeed, the steps in such processes or methods generally may be carried out in various different sequences and orders while still falling within the scope of the present invention. Accordingly, it is intended that the scope of patent protection afforded the present invention is to be defined by the appended claims rather than the description set forth herein.

Additionally, it is important to note that each term used herein refers to that which the Ordinary Artisan would understand such term to mean based on the contextual use of such term herein. To the extent that the meaning of a term used herein—as understood by the Ordinary Artisan based on the contextual use of such term—differs in any way from any

US 7,389,897 B2

5

particular dictionary definition of such term, it is intended that the meaning of the term as understood by the Ordinary Artisan should prevail.

Furthermore, it is important to note that, as used herein, "a" and "an" each generally denotes "at least one," but does not exclude a plurality unless the contextual use dictates otherwise. Thus, reference to "a picnic basket having an apple" describes "a picnic basket having at least one apple" as well as "a picnic basket having apples." In contrast, reference to "a picnic basket having a single apple" describes "a picnic basket having only one apple."

When used herein to join a list of items, "or" denotes "at lease one of the items," but does not exclude a plurality of items of the list. Thus, reference to "a picnic basket having cheese or crackers" describes "a picnic basket having cheese without crackers", "a picnic basket having crackers without cheese," and "a picnic basket having both cheese and crackers." Finally, when used herein to join a list of items, "and" denotes "all of the items of the list." Thus, reference to "a picnic basket having cheese and crackers" describes "a picnic basket having cheese, wherein the picnic basket further has crackers," as well as describes "a picnic basket having crackers, wherein the picnic basket further has cheese."

Turning now to the drawings, a first embodiment of a baby bag 10 that is convertible into a baby carrier is shown in various views in FIGS. 1-6. A first side 12, a first end 14, a top 16, and purse straps 18 are shown in FIG. 1. The top 16 having purse straps is shown in FIG. 2. The top 16 has a pocket 20 attached thereto and a pocket closure 22 such as an area of hook-and-loop fasteners. In one embodiment, the pocket 20 is specifically dimensioned to receive a mobile phone in snug fit therein. In another embodiment, the pocket 20 is specifically dimensioned to receive a baby bottle in snug fit therein. A bottom 24 is shown in FIG. 2A having a pocket 26 attached thereto and a pocket closure 28 such as an area of hook-and-loop fasteners. A second side 30 is shown in FIG. 3. The top 16 is hingedly attached to the second side 30 defining an opening and closing top of the bag for accessing the major interior thereof. The first side 12 is shown in FIG. 4 having a pocket 32 attached thereto and pocket closures 34,36 such as areas of hook-and-loop fasteners. A second end 38 is shown in FIG. 5 having first and second members 40,42 thereof that are joinable by a connector 44 such as a zipper. The first member 40 of the second end 38 is hingedly attached to the first side 12 and the second member 42 is hingedly attached to the second side 30. The first end 14 is shown in FIG. 6 having first and second members 46,48 thereof that are joinable by a connector 50 such as a zipper. The first member 46 of the second end 14 is hingedly attached to the first side 12 and the second member 48 is hingedly attached to the second side 30.

When the first and second members 46,48 of the first end 14 are joined by the connector 50 as shown in FIG. 6, and the first and second members 40,42 of the second end 38 are joined by the connector 44 as shown in FIG. 5, the bag 10 is configured as a generally rectangular bag having multiple outer pockets having pocket closures. The bag is generally carried by purse straps 18. The major interior of the bag is accessible by hinging the top 16 about its attachment to the second side 30. Articles such as diapers and cleaning cloths optionally stored in the various outer pockets are accessible without unfolding the bag and without reaching deeply into the major interior of the bag.

The baby bag is convertible to a baby carrier as shown in various views in FIGS. 7-15. In FIG. 7, the connectors 48,50 are released for separation, respectively, of the first and second members 40,42 of the second end 38 and of the first and second members 46,48 of the first end 14, thereby allowing

6

conversion of the baby bag into the baby carrier by unfolding of the baby bag. The configuration shown in FIG. 1 is referred to nominally herein as a folded configuration for storage and transport of the baby bag. The configuration shown in FIG. 8 is referred to nominally herein as an unfolded configuration defining a baby carrier 10 for receiving a baby. FIG. 7 shows the act of converting the baby bag 10 of FIGS. 1-6 into the baby carrier 10 of FIGS. 8-9.

A base of the baby carrier is defined by the sides and bottom of the baby bag. The first side 12 and second side 30 are each hingedly attached to the bottom 24 such that the baby bag 10 (FIG. 6) is convertible to the baby carrier 10 (FIG. 8) by hinging of the first and second sides about the bottom. Thus, in the context of the baby bag 10 (FIG. 6) the items 12,30, and 24 are referred to nominally herein respectively as the first side 12, second side 30, and bottom 24 of the baby bag 10; and, in the context of the baby carrier 10 (FIG. 8) the same items are referred to as members of the base 52.

As shown in FIG. 8, a collapsible wall 53 is attached to the base 52 and extends around the perimeter thereof. The wall 53 is shown as at least partially collapsed in FIG. 7. When the base 52 is in the unfolded configuration (FIG. 8), the collapsible wall 53 is generally upstanding from the base for encircling a baby received upon the base.

A pocket 55 is attached to the base 52 as show in FIG. 8 and has a gathered elastic opening that opens away from a child received on the base (FIG. 13). The pocket 55 is positioned proximal the legs and under the feet of the child. A blanket, diapers, wipes, or other supplies and articles are optionally stored in the pocket 55.

Also as shown in FIG. 8, a pocket 155 optionally is provided on collapsible wall 53. Pocket 155 is specifically dimensioned to receive the base of an umbrella in snug fit therein for shading of a baby received within a harness 70, which harness discussed in further detail below. The umbrella preferably is lightweight whereby the pocket 155 will support the umbrella in a desired position in order to provide the desired shade.

As shown in FIGS. 9-11, the base 52 of the baby carrier 10 comprises rigid members that are hingedly attached together. A folded configuration of the base 52 is shown in FIG. 10, wherein rigid side members 54,56 are oriented essentially perpendicularly to a rigid central member 58. The rigid side members 54, 56 are each hingedly attached to the rigid central member 58 by flexible hinging members 60 such that overextension beyond an unfolded configuration of the base as shown in FIG. 111 is prevented by the rigid members 54,56, 58. In particular, respective portions 58a,58b of the rigid central member 58 extend beyond axes where the rigid side members 54,56 are hingedly attached to the rigid central member 58. Overextension of the rigid side members beyond the unfolded configuration of FIG. 11 is prevented by abutment of the portions 58a,58b with the rigid side members 54,56 respectively.

In the embodiment shown in FIGS. 9-11, the flexible hinging members 60 comprise flexible portions of fabric that are bonded, adhered, sewn, or otherwise attached to the rigid members of the base while other hinging constructions comprising living hinges and hinges having axles are included in other embodiments. A cushion 59 (FIGS. 10-11) is provided for disposition between the rigid members 54,56,58 and a baby for comfort. Optionally, a washable removable fitted sheet (not shown) or other soft cover receives the cushion and rigid members to retain the cushion in abutment with the rigid members, to protect the cushion from soiling, and to further comfort the baby.

US 7,389,897 B2

7

Another embodiment of an arrangement of rigid members of a base of a baby bag convertible to a baby carrier is partially shown in FIG. **12**. In this embodiment, like that of FIGS. **9-11**, rigid side members **62,64** are each hingedly attached to a rigid central member **66** by flexible hinging members **68** such that overextension beyond an unfolded configuration is prevented by the rigid members **62,64,66**. In this embodiment, however, unlike that of FIGS. **9-11**, respective portions **62a,64a** of the rigid side members extend beyond axes where the rigid side members **62,64** are hingedly attached to the rigid central member **66**. Overextension of the rigid side members is prevented by abutment of the portions **62a,64a** respectively with the rigid central member **66**.

As shown in FIGS. **8** and **13**, a harness **70** is connected to the base **52** of the baby carrier for restraining a baby received upon the base when the base is in the unfolded configuration. The harness is configured to restrain a baby such that the back of the baby is maintained in close proximity to the base. In the embodiment shown in FIGS. **8** and **13**, the harness **70** comprises straps that extend from the base **52**, pass about the shoulders of the child, and connect to a groin area panel that passes between the legs of the child. A breast strap assembly passes across the breast of the child. The harness includes sliding adjusters for providing a comfortable fit about the child and release snaps for conveniently securing and releasing the child.

Furthermore, the baby carrier preferably includes a cushion for partially surrounding and supporting a baby's head when the baby is received on the base. In the embodiment illustrated in FIGS. **8** and **13**, the cushion **72** (FIG. **13**) is removably attached to the base **52** via hook-and-loop fasteners **74** (FIG. **8**). The cushion is provided for comfort and to control the position of the head of the child as may be needed with particular regard to young infants.

A blanket **82** is shown covering much of the child in FIG. **14**. The blanket **82** is optionally removably attached to the base **52** via hook-and-loop fasteners (not shown). Securing pads **84** attached to the blanket are for securing a bottle support **86** (FIG. **15**) to the blanket as shown in FIG. **16**. In the illustrated embodiment, the securing pads **84** (FIG. **14**) comprise hook-and-loop fasteners for engaging corresponding fasteners on the bottom of the bottle support. The bottle support **86** is shown in FIG. **16** with a bottle therein such that the bottle is supported by the support and is retained by a strap **88** (FIG. **15**) of the bottle support **86** such that the child may comfortably feed from the bottle.

A shoulder strap **90** is provided for suspension of the baby carrier **10** therefrom when the baby carrier is in the unfolded configuration as shown in FIGS. **17-18**. The shoulder strap **90** is capable of being removably attached to the baby carrier, for example, by way of releasable clips **9** (FIG. **17**) each located at an opposing end of the baby carrier **10**. A baby received by the baby carrier may be carried by a person utilizing the shoulder strap as shown in FIG. **18**. Toys hanging on a cord may be attached to the shoulder strap, as shown in FIG. **17**, which will swing and move for the amusement of the child as the baby carrier is carried.

Additionally or alternatively, a shoulder strap is removably attached to the baby carrier at opposing sides of the carrier such that the baby carrier may be carried by a person utilizing the shoulder strap, as shown in FIG. **19**. In the arrangement of FIG. **19**, a shoulder strap is attached to the base of the carrier proximal the rigid central member **58** (FIGS. **9,11**) such that, upon lifting the carrier by the strap, biasing of the carrier toward the folded configuration is minimized or avoided. For additional security, a belt worn about the waist of the person

8

utilizing the shoulder strap may be passed through an anchor strap **91** (FIG. **6**) attached to the base of the carrier.

Additionally or alternatively, the baby carrier is capable of being supported by straps **94** as shown in FIG. **20**. The straps **94** are attached to the base of the carrier proximal the rigid central member **58** (FIGS. **9,11**) such that, upon lifting the carrier by the straps as shown in FIG. **20**, biasing of the carrier toward the folded configuration is minimized or avoided.

According to another embodiment of the invention as shown in FIGS. **21-25**, a baby bag **500** (FIG. **23**) is convertible to a baby carrier **500** (FIGS. **21, 24-25**) that comprises a base **502** having an unfolded configuration as shown in FIGS. **21** and **24-25**, and a folded configuration as shown in FIG. **22-23**. The base **502** comprises a first member **502a** for receiving the head and upper torso of a baby, and a second member **502b** for receiving the legs of a baby. A harness **504** is attached to each of the first member and second member for securing a baby when received by the base **502** in the unfolded configuration.

A hinge-control lever **506** extends from the base **502** along the axis along which the member **502a** and second member **502b** are hingedly attached. When the base **502** obtains the folded configuration (FIG. **22**), the first member **502a** abuts the second member **502b**. When the base **502** obtains the unfolded configuration (FIG. **21**), the first member **502a** and second member **502b** are disposed in an essentially coplanar relationship. The first member **502a** and second member **502b** are capable of any relative disposition between the unfolded configuration (FIG. **21**) and the folded configuration (FIG. **22**) when the hinge-control lever **506** is unlocked. For example, in FIG. **24** the first member is slightly inclined from the completely unfolded configuration. Relative movement (hinging) between the first member and second member is arrested when the hinge-control lever is locked.

A first lateral member **508** is hingedly attached to the first member **502a** of the base and a second lateral member **510** is hingedly attached to the second member **502b** of the base. Each lateral member **508,510** includes a number of pockets **512** having optional closure elements such as buckles, zippers, and buttons. The pockets **512** are arranged on the faces of the lateral members that are readily accessible when either the folded configuration (FIG. **23**) or the unfolded configuration (FIG. **21**) of the base **502** is obtained. A pocket **509** attached to the first lateral member **508** may be specifically dimensioned to receive a mobile phone in snug fit therein. Alternatively, pocket **509** may be specifically dimensioned to receive a baby bottle in snug fit therein.

The convertible baby bag **500** is configured as a bag with a bag-carrying strap assembly **514** in FIG. **23**, wherein the first lateral member abuts the first member of the base and the second lateral member **510** abuts the second member of the base. Closure elements **516** attached to the lateral members and comprising straps and quick-release buckles are useful for securing the baby bag in the bag configuration shown in FIG. **23**. The closure elements **516** are further useful for securing the lateral members in rolled configurations as shown in FIG. **24**.

The convertible baby bag **500** includes a collapsible canopy **518** as shown in FIGS. **21** and **24**. The collapsible canopy optionally hinges about the axis along which the first member and second member of the base are hingedly attached.

The convertible baby bag **500** is configured as a baby carrier with a shoulder strap assembly in FIGS. **24** and **25**. The shoulder strap assembly has a single strap portion **520** and a double strap portion **522**. The double strap portion **522** is anchored to the first and second members of the base at

US 7,389,897 B2

9

spaced respective anchor points to assist in stabilizing the baby carrier when carried as shown in FIG. **25**.

A collapsible wall **524** is attached to and extends around the perimeter of the base for encircling a baby received upon the base when the unfolded configuration (FIG. **24**) is obtained. In the illustrated embodiment, the wall **524** includes sections **526** comprising mesh. The wall **524** collapses toward the base when the folded configuration is obtained (FIG. **23**).

Within the convertible baby bag **500** shown in FIGS. **21-25**, a hinging frame **530** as shown in FIGS. **26-28** comprises a first frame member **532** for supporting the first member of the base, a second frame member **534** for supporting the second member of the base, and the hinge control lever **506**. The hinging frame **530** obtains: the unfolded configuration shown in FIG. **26** when the base **502** (FIG. **21**) obtains the unfolded configuration; the folded configuration shown in FIG. **27** when the base **502** (FIG. **22**) obtains the folded configuration; and, an intermediate inclined configuration as shown in FIG. **28** when the base **502** obtains an inclined configuration. Relative movement (hinging) between the first frame member and second frame member is arrested when the hinge-control lever **506** is locked. Construction of the hinging frame **530** prevents overextension of the frame members beyond the unfolded configuration (FIG. **26**).

Another embodiment of a frame for a convertible baby bag is shown in FIGS. **29-30**. In FIG. **29**, the hinging frame **630** includes a first frame member **632**, a second frame member **634**, a canopy support member **636**, and a bracket **638** about which the first frame member, the second frame member and the canopy support member rotate. A leg member **640** is hingedly attached to the first frame member **632** for supporting the first frame member in an inclined configuration as shown. Horizontal support members **642** depend from the bracket providing stability to the hinging frame when placed on a horizontal surface and preventing overextension of the first frame member and second frame member beyond the horizontal. The hinging frame **630** is shown in FIG. **29** in an inclined configuration that is intermediate a folded configuration (not shown), wherein the first and second frame members are vertically disposed, and an unfolded configuration (not shown), wherein the first and second frame members are horizontally disposed.

As shown in FIG. **30**, wherein the first and second frame members are maintained as in FIG. **29** but wherein the bracket **638** is not shown, overextension of the first and second frame members **632,634** beyond a horizontal configuration is further prevented by respective abutting surfaces **644,646** that abut each other when the horizontal configuration is obtained.

In addition to the foregoing detailed description, it further is noted that the embodiment of the baby carrier as shown, for example, in FIG. **21** includes feet **555** for elevation of the base of the baby carrier above a surface upon which the baby carrier may be placed. Elevation of the baby carrier is advantageous for insulating the baby carrier from heat exchange with the support surface. Feet **556** further are provided in the frame of, for example, FIG. **26** for similarly elevating a base of a baby carrier. Moreover, the frame of FIG. **26** further defines openings **557** extending therethrough containing air for further insulating the base from any support surface upon which the base may be placed. Similarly, the frame of the embodiment of, for example, FIG. **30** also includes feet **558** and openings **559** separated by support walls for elevating and insulating a base of a baby carrier.

What is claimed is:

1. A baby bag that is convertible to a baby carrier, comprising:
   (a) a base that is movable between,

10

   (i) a folded configuration defining a baby bag for storage and transport thereof, and
   (ii) an unfolded configuration defining a baby carrier for receiving a baby thereon; and
 (b) a harness connected to said base for restraining a baby received upon said base when said base is in said unfolded configuration, said harness configured to restrain a baby such that a back of the baby is maintained in close proximity to said base;
 (c) wherein said base includes first and second rigid planar members coupled together and movable,
   (i) from a first position, corresponding to said base being in the unfolded configuration,
     (A) wherein said first and second rigid members are disposed in coplanar relation to one another, with proximal ends thereof spaced apart from one another, and
     (B) wherein said first and second rigid members are configured to support a baby received upon said base,
   (ii) to a second position, corresponding to said base being in the folded configuration,
     (A) wherein said first and second rigid members are disposed in spaced parallel relation to one another, and
     (B) wherein said first and second rigid members define therebetween an interior space;
 (d) wherein said base comprises a third member that is rigid and planar and that extends between and connects said first rigid member and said second rigid member; and
 (e) wherein overextension of said base is prevented by abutment of an overlapping portion of said first rigid member with said third rigid member, and abutment of an overlapping portion of said second rigid member with said third rigid member.

2. The baby bag of claim **1**, wherein the moving of said base from the folded configuration to the unfolded configuration includes rotating said first rigid member relative to said third member, and rotating said second rigid member relative to said third member.

3. The baby bag of claim **1**, wherein said first and second rigid members are parallel to, but not coplanar with, said third rigid member when said first and second rigid members are in their said respective abutment with said third rigid member for preventing overextension of said base.

4. The baby bag of claim **1**, further comprising a shoulder strap removably attached to said base proximate said third member.

5. The baby bag of claim **1**, wherein said base is cushioned for receiving a baby in comfort.

6. The baby bag of claim **1**, wherein said baby bag includes a plurality of pockets, and wherein at least one said pocket is disposed on a member that is connected to said base for hinging movement relative thereto.

7. The baby bag of claim **1**, wherein said baby bag includes a plurality of pockets, and wherein at least one said pocket is dimensioned to receive a mobile phone.

8. The baby bag of claim **1**, wherein said baby bag includes a plurality of pockets, and wherein at least one said pocket is dimensioned to receive a baby bottle, said pocket being disposed on the outside of said baby bag when said base is in said folded configuration.

9. The baby bag of claim **1**, wherein said baby bag includes a plurality of pockets, and wherein at least one said pocket is dimensioned to receive an umbrella when said base is in said unfolded configuration for shading of a baby received upon said base.

US 7,389,897 B2

11

10. The baby bag of claim **1**, wherein said baby bag further comprises a canopy for covering of a baby when received on said base.

11. The baby bag of claim **1**, further comprising an anchor strap connected to said base proximate said third member for receipt there through of a belt.

12. A method of using the convertible baby bag of claim **1** for carrying a baby, comprising the steps of,

  (a) converting the baby bag to the baby carrier by moving the base into the unfolded configuration;

  (b) placing a baby upon the base in the unfolded configuration;

  (c) restraining the baby to the base using the harness such that a back of the baby is maintained adjacent the base; and

  (d) carrying the base with the baby restrained to the base.

13. The baby bag of claim **1**, wherein said baby bag further comprises a collapsible wall attached to and extending around a perimeter of said base for encircling a baby received upon said base when said base is in said unfolded configuration, said wall collapsing toward said base when said base is moved into said folded configuration.

14. The baby bag of claim **13**, wherein said collapsible wall includes sections comprising mesh.

15. The baby bag of claim **1**, wherein said baby bag further comprises a cushion for partially surrounding a baby's head when received upon said base.

16. The baby bag of claim **15**, wherein said cushion is removably attached to said base via hook-and-loop fasteners.

17. The baby bag of claim **1**, wherein said baby bag further comprises a blanket for covering of a baby received upon said base.

18. The baby bag of claim **17**, wherein said blanket is removably attached to said base via hook-and-loop fasteners.

19. A baby bag that is convertible to a baby carrier, comprising:

  (a) a base that is movable between

    (i) a folded configuration for storage and transport of the baby bag, and

    (ii) an unfolded configuration for receiving a baby; and

12

  (b) a shoulder strap attached to said base when said base is in said unfolded configuration for suspension of said base, whereby a baby received upon said base may be carried by a person utilizing said shoulder strap;

  (c) wherein said base includes first and second rigid planar members coupled together and movable,

    (i) from a first position, corresponding to said base being in the unfolded configuration,

      (A) wherein said first and second rigid members are disposed in coplanar relation to one another, with proximal ends thereof spaced apart from one another, and

      (B) wherein said first and second rigid members are configured to support a baby received upon said base,

    (ii) to a second position, corresponding to said base being in the folded configuration,

      (A) wherein said first and second rigid members are disposed in spaced parallel relation to one another, and

      (B) wherein said first and second rigid members define therebetween an interior space;

  (d) wherein said base comprises a third member that is rigid and planar and that extends between and connects said first rigid member and said second rigid member; and

  (e) wherein overextension of said base is prevented by abutment of an overlapping portion of said first rigid member with said third rigid member, and abutment of an overlapping portion of said second rigid member with said third rigid member.

20. A method of using the convertible baby bag of claim **19** for carrying a baby, comprising the steps of,

  (a) converting the baby bag to the baby carrier by moving the base into the unfolded configuration;

  (b) placing a baby upon the base in the unfolded configuration; and

  (c) suspending the base and the baby placed thereon by the shoulder strap.

*   *   *   *   *

he following is intended as a partial list of terms and definitions you might run across during your inventing process. The definitions here are just quick summaries to acquaint you with these terms, and you should not rely on the legal terms as complete explanations of legal requirements. You should not rely on these definitions in drafting documents (especially patent applications) or in negotiating agreements. Please read the chapter in which the terms are actually covered to see the context in which we are discussing them.

abandonment: Circumstances in which rights in a patent application or an invention have been lost either by intent of the inventor or through unintentional action such as a failure on the part of the inventor to file a proper reply to the USPTO within a specified time period. Abandonment status may be withdrawn or an abandoned application revived in certain instances (mailing errors, failure to pay the issue fee) if it was unintentional. Abandonment of an invention must be intentional.

angel investor (or "business angel" in Europe): A high-net-worth, affluent individual who provides capital for a business startup, usually in exchange for convertible debt or ownership equity. A small but increasing number of angel investors organize themselves into **angel groups** or **angel networks** to share research and pool their investment capital.

assignment: In relation to patents and trademarks, the transfer of ownership of intellectual property from the inventor to a third party.

big-box retailer: Large "category killer" retailers who specialize in certain industries such as home improvement (Lowe's, Home Depot), office supplies (Staples, Office Depot), crafts (Michaels, Hobby Lobby).

bootstrapping: A means of advancing oneself or one's business by minimizing expenses and doing as much as possible with the fewest financial resources.

brand: A name used to indicate the maker of a product or service and to differentiate it from others.

burden of proof: The obligation to establish a contention as fact by evoking evidence of its probable truth.

claims: The part of a patent wherein the invention and the aspects of the invention that are legally enforceable are defined.

continuation/continuation-in-part application: An application filed after an original patent application in order to amend the original in some way.

contributory infringement: When somebody knows of the direct infringement of another and substantially participates in that infringement, such as inducing, causing, or materially contributing to the infringing conduct.

copyright: The exclusive right to make copies, license, and otherwise exploit a literary, musical, or artistic work, whether printed, audio, video, used to protect original works of art (books, movies, pictures, songs, paintings). Works granted such right by law on or after January 1, 1978, are protected for the lifetime of the author or creator and for a period of 50 years after his or her death.

cost-justify: To justify the allotment or spending of a specific sum of money for an investment or procurement.

critical date: The date by which an application must be filed to avoid losing rights in the invention.

declaration: See oath.

design patent: A patent that protects only the ornamental appearance of an article.

direct infringement: See infringement.

direct response television (DRTV): Any TV advertising that asks consumers to respond directly to the company, usually by calling an 800 number or by visiting a website.

direct sales methods: A method of selling in which the seller bypasses traditional stores and sells directly to the end user (such as Avon, Amway, Tupperware parties).

distribution channels: Used to describe the various ways of getting a product from the manufacturer to the end user, which include retail (mass merchants, big-box retailers, specialty retailers, catalogs, TV shopping) as well as wholesalers, distributors, brokers, and other forms of selling.

doctrine of equivalents: A judicial doctrine that says a patent holder is entitled to protection from infringers who make trivial modifications to a product that has the same function as the holder's invention and is not literally infringing on that invention.

embodiment: Giving concrete form to an abstract concept; also, that concrete form.

examiner: The representative at the USPTO who will review the patentability of an application. This person is not necessarily an attorney but has passed the patent bar and has some technical background.

focus group: A marketing research tool in which a small group of people (about eight to ten individuals) engages in a roundtable discussion of selected topics of interest in an informal setting. The discussion is typically directed by a moderator who guides the discussion in order to obtain the group's opinions about or reactions to specific products or marketing-oriented issues known as test concepts.

gross margin: The percentage of profit generated from sales determined by dividing the gross profit (revenue minus the cost of goods sold) by total revenue.

gross profit: Dollar profit determined by subtracting cost of goods sold (COGS) from total revenue (sales).

independent inventor: An individual who creates something new and useful on his or her own. Also called an individual inventor.

information disclosure statement (IDS): A list of all patents, publications, U.S. applications, or other information submitted for consideration by the office in a nonprovisional patent application filed under 35 U.S.C. III(a) to comply with applicant's duty to submit information which is material to patentability of the invention.

infringement: Making, selling, or using an invention, trademark, or other exclusive intellectual property without the legal right to do so.

initial public offering (IPO): A corporation's first offer to sell stock to the public.

intellectual property: Property that results from original creative thought; also, rights belonging to this property, such as patents, copyrights, trademarks, and trade secrets.

invention: A new, useful process, machine, improvement, or other innovation that did not exist previously and that is recognized as the product of some unique intuition or genius, as distinguished from ordinary mechanical skill or craftsmanship.

invention promotion company: A company that offers to represent inventors in licensing their ideas to other companies.

inventor's notebook: A collection of sequentially numbered and permanently bound pages (that prevents someone from adding or removing pages) in which to record invention notes. All notes should be signed by a witness (who will not benefit financially from the invention), which can establish the first date of invention. Also called a lab notebook.

letter of credit: A bank instrument that is issued to a vendor, allowing the vendor to draw from the bank a specified dollar amount in the event payment is not made per the terms of an agreement.

letter of intent: A written statement expressing the intention of the undersigned parties to enter into a formal agreement, especially a business arrangement or transaction.

licensee: A person, company, or other entity to whom a license is granted.

licensing agreement: A contract giving someone the legal right to use a patent, trademark, copyrighted materials, or other intellectual property for a specific duration, purpose, and fee.

licensor: A person, company, or other entity who owns the rights in the property that is being licensed.

market research: The gathering and studying of data relating to consumer preferences, purchasing power, et cetera. Normally done prior to introducing a product in the market.

market share: The specific percentage of total industry sales achieved by a single company's particular product in a given period of time.

market traction: Describes a condition in which a product is predictably and/ or consistently sold to buyers in the same market segment or population. A company has attained market traction when its product is predictably sold to buyers in the same market segment.

marketability: The probability of selling property at a specific time, price, and terms.

mass merchant: A large, multi-unit chain retailer such as Wal-Mart, Target, and Kmart.

new matter: New information added to a patent application or patent.

niche market/target market: A specialized and profitable part of a commercial market.

nondisclosure agreement (NDA): A contract whereby one promises to treat information confidentially and not give out information without proper authorization.

nonobvious: A criterion in Section 103 of the U.S. Code patent statutes which requires that an invention cannot receive a valid patent if the invention could be readily deduced from publicly available information by a person of ordinary skill in the art.

nonprovisional patent application: An application for a patent to be filed (i.e. utility, design, plant, and reissue). This term includes all types of applications except provisional. The nonprovisional application establishes the filing date and initiates the examination process, whereas the nonprovisional utility patent application must include specifications including a claim or claims, drawings when necessary, an oath, and the prescribed filing fee.

novelty: An idea that is unique, new, and not known to exist anywhere in the world.

oath: Also called a declaration. A statement complying with the laws of the state or country where made in which an applicant for a patent declares by document that 1) he or she is the original or sole inventor, 2) which state of which country he or she is a citizen, 3) that he or she has reviewed and understands the content of the specifications and claims the declaration refers to, and 4) acknowledges the duty to declare information that is material to patentability.

office action: A letter in which a patent examiner allows, rejects, or objects to an application and explains his or her reasons for doing so. The patent applicant has a set period in which to respond.

on-sale bar: In the United States, selling an invention or offering the invention for sale presents a bar to the patentability of the invention beginning one year after the date of sale or offer for sale. In almost every other jurisdiction, selling an invention or offering the invention for sale presents an immediate bar to patenting the invention.

patent: The exclusive right granted by a government to an inventor to prevent others from making, using, or selling an invention for a certain number of years.

patent attorney: An attorney who must have special training and pass a special bar examination that requires him or her to have either a science or engineering degree, special lab experience, or knowledge deemed equivalent.

performance metrics: Ways to identify and evaluate the success of a product which could include sales, returns, inquiries, response rate, Internet traffic, et cetera.

preproduction samples: The first few exemplars of a product created prior to the final production run.

presentation ("looks-like") prototype: A prototype that resembles the finished product but does not necessarily function, or "work like," the final product will. This is used primarily to show the appearance of the product to potential buyers to solicit feedback and/or orders.

product development: The process of going from an idea to a finished product. Also, the improvement of an existing product or the cultivation of new kinds of products.

product fulfillment company: A company that provides individuals or companies with a support structure to handle all of their back-end operations, including warehousing, order picking, packing, and shipping services.

product liability insurance: Coverage that protects an individual or firm against financial loss when a person files suit claiming they or their property were harmed by its product.

profitability: The point at which the revenue exceeds the cost of goods and general operating expenses.

proof of concept ("works-like") prototype: A physical demonstration that a business model or product is feasible. It may not resemble the final product, but functions similarly.

prototype: A physical representation of an idea that allows an individual to observe and interact with it.

provisional patent application: Allows filing of a U.S. national application without a formal patent claim, oath, or any information disclosure (prior art) statement. It provides the means to establish an early effective filing date in a nonprovisional patent and automatically becomes abandoned after one year. It also allows the term "patent pending" designation to be used. The USPTO does not examine provisional applicatons.

public domain: Property rights unprotected by copyright or patent that are subject to appropriation by the community at large.

purchase: To acquire by the payment of money or its equivalent; buy; also, an item that has been so obtained.

purchase order: A commercial document used to request someone to supply items in return for payment and providing specifications and quantities.

quality control standards: Standards for establishing policies and procedures that provide reasonable assurance that all of a firm's processes and products are conducted in accordance with applicable professional standards.

rapid prototyping: The automated construction of physical objects using solid free-form fabrication. Rapid prototyping takes virtual designs from computer-aided design (CAD) or animation modeling software, transforms them into thin, virtual, horizontal cross sections, and then creates each cross section in physical space, one after the next, until the model is finished.

receiving office reexamination: After issuance of a patent, a new examination that takes place upon a request from a third party that the USPTO take another look at the patent. The USPTO does not get involved unless an examiner decides a substantial question of patentability exists. On the reexamination the USPTO can consider prior art it did not consider in the first examination.

retailer: A person or company that sells a product to the consumer.

revenue: The income received from the sale of a product or service.

reverse engineering: Study or analysis in order to learn details of design, construction, and operation, perhaps to produce a copy or an improved version; also, the use of data gathered in this manner.

right to exclude: The right of a patent holder to prevent others from making, practicing, importing, or selling their invention.

royalty: A compensation or portion of the proceeds paid to the owner of a right, patent, or trademark, for the use of the item so protected.

Small Business Administration (SBA): A federal agency, created in 1953, that among other things grants or guarantees long-term loans to small businesses.

supply chain: The network of retailers, distributors, transporters, storage facilities, and suppliers that participate in the sale, delivery, and production of a particular product.

survey: A series of questions used in market research to define and assess a product or service.

trade dress: The distinctive packaging or design of a product that promotes the product and distinguishes it from other products in the marketplace.

trade show: A gathering or convention at which companies in a particular industry gather to demonstrate or promote their products.

trademark: Any name, symbol, figure, letter, word, or mark adopted and used by a manufacturer or merchant in order to distinguish his or her goods from those manufactured or sold by others. A trademark is a proprietary term usually registered with the Patent and Trademark Office to assure exclusive use by its owner.

underwrite: To review and take action through a funding means.

United States Patent and Trademark Office (USPTO): The governmental agency charged with determining the patentability of all United States applications, issuing all U.S. patents and trademarks, maintaining the government's patent database, and acting as a receiving office for some international applications.

value proposition: The amount paid for a product or service and what a customer receives for that value.

venture capital (VC): A type of private equity capital typically provided to immature, high-potential, growing companies in the interest of generating a return through an eventual liquidity event such as an IPO or sale of the company. Venture capital investments are generally made as cash in exchange for shares in the invested company.

Acknowledgments

As I have said many times before, an idea without execution is just an idea. I have learned that this applies to writing a book as well. For years I have talked about doing it, I even told a few friends and colleagues about it, but I never found the time to pursue it. Taking a bit of my own advice, I finally found the time and "did it." As simple as that sounds, it was far from it.

I tell inventors that they need to surround themselves with people who have "done it before" and people who can assist them in the process. This book is the result of just that. Many people contributed to the development and were supportive during the process. This is a perfect opportunity to recognize and thank everyone for their involvement. This would not be a complete guide without the help and great advice from my coauthor, Jill. Her knowledge and passion for the inventor community was evident from the very first time I met her.

Once the "proof of concept" was created, it was my literary agent, David Fugate, who pitched the prototype and found a home for the book at Workman Publishing. Thank you Peter Workman for seeing the potential in this project and making it happen. I also thank our editor, Megan Nicolay (a closet inventor herself), for holding my hand through the process, and the rest of the Workman team for working so tirelessly on this book.

My partners and coworkers at Enventys, *Inventors Digest, Everyday Edisons,* and Edison Nation are the ones that make the dreams come true. Without their hard work and daily execution, it would just be an idea. Their dedication and passion for the business and each one of the products that we develop is incredible. It is inspiring to work with these individuals and watch as they apply their skills, as well as how they interact with inventors.

I cannot forget the inventors. Over the years, I have met tens of thousands of you—at casting calls, inventor events, and online. Your great ideas continue to amaze me. Don't stop innovating.

None of this would be possible without the support and encouragement of my family. My parents, who supported my decision to become an entrepreneur (and not the doctor they initially dreamed I would be), and my children, Hailey and Aaron, who I miss terribly every time I get on a plane and leave town (which is extremely frequently). My kids represent the future of innovation and I hope that they will be inspired to follow through with their dreams and never settle for less. We need to encourage innovation, teach it, and celebrate the achievements. Finally, I need to thank my wife, Liz, who holds it all together while I am gone. Without her love and support, I wouldn't be the person I am today. She has been with me from the very first fraternity room company until today, and I cannot imagine anyone else I would want to make this journey with.

Good luck to all in your own journey.—LJF

Thank you to Louis for taking my call out of the blue, looking at my draft chapter, and putting the weight of a veritable Edison Nation behind it. Special thanks to David Fugate for finding this book a great home and a voice at Workman.

I also want to say thanks to the Wisconsin Department of Commerce for seeding the first state-supported inventors club program in the nation, based on the innovative model for these clubs set by Terry Whipple, director of economic development in Juneau County, and to Susan Noble who helped me start my first club. We all thank the Wisconsin Entrepreneur's Network for giving us guidance in the form of Dave Linz, Southeastern Regional Director, and my good friend Dan Steininger for founding Milwaukee Biz Starts. All of these people taught me the things I never learned in law school about taking ideas to market.

Thanks most of all to Megan Nicolay, our editor, for making all of the ideas mesh, and to my proofreading attorney spouse, Dan, with the red pen and reading glasses on the nightstand.

You are all true friends of inventors everywhere.—JGW

Index

CREDITS

Cover illustration by Hanoch Piven/www.heflinreps.com

PHOTO CREDITS: John Abdo: p. 156; **Alamy Images:** mediablitzimages (uk) Limited p. 8. rostislav foursa p. 68, dk p. 125; **AP Images:** Ben Margot p. 170, Sal Veder p. 177; **Barbara Carey:** p. 173; Corbis: Bettmann p. 146; **Courtesy Edison Nation:** p. 3, 18, 20, 30, 31, 35, 45, 54, 85, 124, 138, 148, 207 (right); **Getty Images:** p. 111, Luke Frazza/AFP p. 130; Tim Boyle p. 128, 142, Time and Life Pictures p. 38, 147, 188, Derek Hudson p. 57, Hulton Archive p. 4, 49, Martin Schreiber p. 66, French School/The Bridgman Art Library p. 86; **Prestige Photo:** p. 207 (left); **SkyMall, Inc:** p. 132; **Superstock:** age fotostock p. 26 (center); Digital Vision Ltd. P. 189; Dynamic Graphics Value p. 26 (left); Ingram Publishing p. 198, Charles Orrico p. 183; **United States Patent and Trademark Office:** August Bartholdi p. 101, G. D. Beauchamp p. 9, James Dyson p. 58, E. H. Land p. 190, Erwine and Estelle Laverne p. 102 (top), Philip Leder and Timothy A. Stewart p. 112, J. H. Lemelson p. 109, James M. Lowrance and Johnny W. Hall p. 152, Karen Madigan and Lynn McIntyre p. 55, Saul Palder p. 81, Maria P. Pistiolis, Louis J. Foreman, Daniel L. Bizzell and Ian D. Kovacevich p. 213-227, Eero Saarinen p. 102 (bottom), John Suckow p. 13, E. S. Tupper p. 48, E. C. Walker p. 98